W9-COF-565

don't be trashy

don't be trashy

A Practical Guide to Living
with Less Waste and More Joy

TARA McKENNA

RODALE.

New York

Copyright © 2022 by Tara McKenna

All rights reserved.
Published in the United States by Rodale Books, an imprint of Random House,
a division of Penguin Random House LLC, New York.
rodalebooks.com

RODALE and the Plant colophon are registered trademarks of
Penguin Random House LLC.

Library of Congress Cataloging-in-Publication Data
Names: McKenna, Tara, author.
Title: Don't be trashy : a practical guide to living with less waste and
more joy / Tara McKenna.
Other titles: Do not be trashy
Description: First edition. | New York : Rodale, [2021] | Includes
bibliographical references and index.
Identifiers: LCCN 2021008488 (print) | LCCN 2021008489 (ebook) |
ISBN 9780593232965 (trade paperback) | ISBN 9780593232972 (ebook)
Subjects: LCSH: Waste minimization—Popular works. | Sustainable
living—Popular works. | Recycling (Waste, etc.)—Popular works. |
Refuse and refuse disposal—Popular works.
Classification: LCC TD793.9 .M37 2021 (print) | LCC TD793.9 (ebook) |
DDC 363.72/8—dc23
LC record available at https://lccn.loc.gov/2021008488
LC ebook record available at https://lccn.loc.gov/2021008489

ISBN 978-0-593-23296-5
Ebook ISBN 978-0-593-23297-2

Printed in the United States of America

Book design by Andrea Lau
Cover design: Anna Bauer Carr and Kat Lynch
Cover photographs: Fernando Valencia/Addictive Creative/Offset

10 9 8 7 6 5 4 3 2 1

First Edition

To future generations, you will inherit the earth as we leave it for you.

Contents

Contents

Preface

It Started in Bali

It was March 2013, and I was headed to Indonesia for my dad's wedding. There was already a lot going on in my life; I was transitioning into my first full-time career job as an urban planner and moving cities for the position, plus I was planning my own wedding. My then fiancé, now husband, was traveling through Asia and we met in Jakarta, for my dad's wedding. It all felt incredibly overwhelming because my sister and I hadn't yet met my dad's beautiful Indonesian wife-to-be or her children, and our two families were about to come together.

Their wedding at the Shangri-La Hotel was undeniably picturesque and extravagant. The cuisine alone was a world tour for my taste buds, ranging from fresh sushi to spicy Indian curry and build-your-own desserts. Everyone was dressed to the nines. I wore a full-length gold gown with my blond hair in an updo and makeup that was red carpet ready. For their ceremony, my dad and his wife wore traditional Indonesian attire that was beautifully embellished and elaborate. Afterward, he changed into a tuxedo and his wife wore what I can only describe as a princess dress from a Disney movie: It was a beautiful ball gown that caught the attention of all their family and friends. After their wedding in Jakarta, the plan was to head to Bali as a big, happy family so all of us could get to know one another better while we enjoyed the panoramic views. Little did I know that this trip to Bali was about to change my life forever—and in more ways than one.

We explored Jakarta for a few days, then flew to Bali, a place idealized as a relaxing, pristine island getaway with vibrant coral reefs, unspoiled crystal-clear beaches, rows of rice paddies, active volcanoes, and old-growth rain forests. Before we'd even bought our plane tickets, I knew I wanted to snorkel at a place like the Blue Lagoon. I had seen photos and videos of the bright-blue waters and coral reefs bursting with sea life. But when we finally went snorkeling and I dove below the surface, I found myself watching the ocean currents gently sway with trash and discarded plastics suspended among the innocent fish and colorful corals. At the time, I had no idea that Indonesia is one of the top five countries worldwide known for dumping plastic into our oceans, alongside Vietnam, China, Thailand, and the Philippines, but I knew something needed to change.[1] What I've since learned is that much of the trash being dumped out of Southeast Asia has actually been imported from other parts of the world![2]

Perhaps naively, I wanted the entire world to know that our lifestyles were ruining these pristine destinations. My gut reaction was that I needed to go back to school to become a marine conservationist—that this was the only way I would actually be able to make a difference (keep in mind, I'd already spent six-plus years in university). I figured that I should be an expert, but looking back, that wasn't the most practical response. No one needs a special degree to support conservation, and if I had pursued that special degree, I might have deferred my decision to begin making concrete changes to my life that would help the environment.

At the same time, I couldn't stop talking about the trash issue with my family while we were on vacation. They were definitely concerned, but I took it to a different level by pondering these types of questions aloud to anyone who would listen (voluntarily or not): I wonder how the trash is getting into the sea. Is it litter? Is it illegal dumping? How is industry involved? What can we do about it? Should it be tackled by the government, by companies, or by individuals like me? All of these questions and more were swirling in my head and flowing freely out of my mouth like water from an open faucet. Pummeling my family with all of these inquiries while we were on vacation may not have been the best approach for getting them to care as much about the trash as I did, but during that trip, a fire grew inside me, and I knew it was there to stay. I had always wanted to have a positive impact on the planet and inspire change in other people, and seeing the state of things in Bali was the fuel I needed to kick-start my journey.

Less than twenty-four hours later, all of those dreams came crashing down. While I was losing sleep over the great Pacific garbage patch, back in Canada one of my best friends, Paige, had been desperately trying to get ahold of me. At this point, I was pretty much off the grid with my fiancé in Bali—we had parted ways with my family to travel, just the two of us—and checking my phone wasn't on the agenda. Eventually one of her messages got through. This was followed by many more. Reading the string of messages, I saw that they were becoming increasingly urgent, so I called her as soon as I could. The Wi-Fi in our low-budget hotel room was shoddy, so I rushed to the main office for a better signal. While I ran down what seemed like a million stairs to get to the office, I thought about Paige's parents. *Is everything okay with them?* I wondered to myself. Maybe they've had a health scare, or maybe Paige had an emergency in the family. These were all assumptions, however, because her messages to me were urgent yet vague, and for a good reason.

Through a slightly broken long-distance connection she told me to sit down. *Shit,* I thought. *What the heck is going on? This cannot be good.* Given the time difference between Bali and Toronto, I knew it was an ungodly hour for Paige to be trying to reach me. Aside from that, it's hard to remember how the conversation really went down. It's all a blur in my mind now. The news was worse than I ever could have imagined. Our other best friend had just died in a car crash. She was only twenty-six. My fiancé and I took the next flight home.

It took at least one year for me to pull myself out of a deep, dark, and lonely place of grief, and unfortunately, the next few years were sprinkled with more loss. It's important to take time to recover from the devastating curveballs life throws at you, and I don't regret the detour my life took to heal. Even though my motivation to pursue anything took a back seat for a long time, I hadn't forgotten the awakening I'd experienced in Bali.

Through my grief I spent a lot of time walking in and soaking up nature. Forest bathing, as it's known in English—translated from *shinrin-yoku* in Japanese—is a Japanese tradition that has become a health trend in North America that is essentially therapy from the wilderness (I strongly recommend it!). My time in the woods was incredibly healing. With the leaves rustling in the wind and the calmness of birds fluttering by, time stood still while I was lost in my own thoughts. These moments taught me to have an appreciation for my life so far and to treasure the joy of my memories with loved ones who have passed on. Being in nature also reminded me of how closely tied to the natural world humans are, yet as a society we often don't behave that way. Taking the time to enjoy nature reinspired my desire to get back into sustainability.

With time and healing, I became gradually more energized and began to revisit my passion for environmental issues and the state of the world. I watched documentaries about the environment like *Planet Earth, Cowspiracy, Food, Inc., A Plastic Ocean, The True Cost, Before the Flood,* and many others. I read books from the library—so many books! *The Story of Stuff* by Annie Leonard, *Cradle to Cradle* by William McDonough and Michael Braungart, *Overdressed* by Elizabeth L. Cline, *The Zero-Waste Lifestyle* by Amy Korst, and *This Changes Everything* by Naomi Klein, among others. The more information I consumed, the more I felt reconnected with the epic and life-changing moment I had experienced in Bali. Plus, each of these documentaries and books, along with other sources of information from blogs, community events, and inspirational environmentalists (think David Suzuki, Jane Goodall, and Sir David Attenborough), motivated me to adopt lifestyle changes that were more conscious of the environment. These changes revolved around reducing my waste—more on that shortly.

As I came to accept that I had to go on living my life even though it was different now, I realized something important: There are some things in life that we have no control over but many that we do. For instance, I don't have any control over how much chocolate is out there in the world (a lot of it) tempting me with its sugary goodness, but I do have control over how much of it I choose to eat. When it comes to climate change and the state of our environment, you'll often hear people say that the responsibility to reduce carbon emissions or eliminate plastic waste or avoid ecosystem collapse shouldn't rest on the shoulders of individuals; instead, our governments should be leading the charge, metal straws in one hand and save-the-environment-from-catastrophe legislation in the other. And while I don't disagree that governments play a critical role in shaping our planet's future (they absolutely do!), my question to the critics is this: *Have you ever actually tried to get your driver's license renewed? Or dealt with the government on any other type of application?* It's like watching paint dry; it can take *forever*.

The point is, dealing with the government can be a long, complicated and bureaucratic process. And that's just to accomplish one seemingly simple task for a single person. So if we're waiting for someone else to get their shit together (and let's be frank here; not everyone even agrees on what it looks like to get our shit together for the planet), we could be waiting a long time. I'm not saying we should give the people in charge a free pass on this one—civic engagement is *essential,* so definitely put the pressure on—but I do think we can and should be the ones to steer the conversation and demand that we get there quicker with our actions. Take disposable plastic straws as an example. Have you noticed that since ocean plastic has become a global concern, and in combination with tons of individuals refusing plastic straws, they're becoming a thing of the past? Yep, we have that kind of power. It's time to embrace it and make some (plastic-free) waves.

Obviously, you picked up this book because you already want to do your part to ensure we don't all die in a fiery hellscape, which is why I won't dwell on all of the reasons it's important (a quick scan of the news makes them quite clear). The thing is, our individual actions matter. And no, I don't mean like the anti-litter campaigns that large multinational companies (notorious polluters) organize, which put the onus of the world's plastic pollution problems on individuals; that's not cool. Still, individual actions make a difference in the world around us, especially when they become global movements. Better yet, it's these global movements that put pressure on governments and large corporations to address important issues like plastic waste and climate change. Take Greta Thunberg, for example. Greta's seemingly simple and lonely climate strike from school was a solo activity in 2018 but became a global movement just a year later, and soon Greta was meeting with world leaders to discuss ways to combat climate change. What am I getting at here? Individual actions can create monumental, collective impact! Don't you want to get in on that? I do! While this book focuses on individual lifestyle changes (like refusing plastic straws), it's important to reflect on how these choices fit within the bigger picture (like when fast-food chains get the message and ditch plastic straws—yassss). This isn't an either/ or scenario; we are in this together!

What I didn't quite realize when I embarked on my personal zero-waste journey was all of the ways that reducing my waste would simplify and enhance my life. The past few years have actually resulted in a full-on transformation. More than ever before, I'm purposeful about most of my actions (I'm not perfect, so I'll stick to the word *most*) and my purchases. Side benefits from my more sustainable and low-waste lifestyle have included eating healthier; meeting new people and making new friends; getting involved in my community (civic engagement!); learning to budget better; making calculated purchases free from buyer's remorse; simplifying and decluttering my home; being empowered by making mindful decisions; feeling grateful for having the opportunity to inspire others to live more sustainably (even if it's imperfect); and so much more. Everything in my life feels more streamlined and intentional, and that in and of itself brings me joy.

Mostly? I feel like a total badass. I'm going against the grain, but sustainable living is slowly becoming more common and will eventually be the norm. My hope is that as you read this book, you'll be inspired to transform your life and become less trashy, so that you'll experience many of these benefits, too. Oh, and save the world while you're at it—you badass, you!

No one could have predicted that my best friend would die in a freak accident, and there's nothing I could have done to change her fate. But that isn't true of everything in life. There were things I could do—especially when it came to the environment—that would actually make a difference. In a roundabout way, losing my best friend actually made me even more determined to act—and to act in a much more deliberate way than I might have otherwise.

Experiencing how precarious life on this earth can be left me no choice but to do something to preserve it, with an emphasis on reducing my waste and becoming less trashy. I don't mean to get all sentimental about this, but here's something that I really do believe: In the same way that one moment has the power to end a young person's life, so too does one decision have the power to change the world.

One small decision inspires another, then another, and before you know it, millions of people have decided that they don't need to use plastic straws and other single-use plastics. I am also hopeful that governments will not only start to catch on but catch up, sparking a societal paradigm shift. Maybe these goals sound lofty, but someone's gotta have them! I am optimistic that this book will help you discover how easy it can be to make a few small changes to your life that will make a big difference in the way your actions affect the world. You'll become less trashy, one step at a time!

Your Don't Be Trashy Challenge

Are you ready to jump into your Don't Be Trashy Challenge? Here's the scoop. Each chapter of the book centers around a specific theme related to low-waste living—with a total of twelve chapters—making it easy to implement month by month over the course of a year. As you tackle a new aspect of low-waste living each month, you'll build the momentum of regularly taking small and practical steps to make some big changes over the course of a year. If the month-by-month timeline doesn't feel right to you, feel free to speed it up or slow it down.

My recommendation is that you read the whole book from cover to cover, then go back and start your journey at chapter 1. Regardless of how you make your way through the book, I can guarantee you will see results if you *actually put in the work.* Need a mantra as you go? Just remember these words each time you make a choice: "Don't be trashy." I'll provide lots of ideas for *how* you can start reducing your waste, but when it comes down to it, this project is as simple as this: making the lower-waste choice when possible. Sometimes you will be trashy and sometimes you won't, but this book will help you move toward the less trashy end of the spectrum if you make the changes that serve you and your lifestyle. The outcome? A less wasteful, more intentional life with more joy!

Here are the nuts and bolts: In chapters 1 through 3 I'm going to help you get your shit together.

This section is super important because it's the foundation that sets up your whole year for success. It's about revisiting what it *really* means to reduce, reuse, and recycle (in chapter 1). You'll define your low-waste goals and find out why you want to embark on this journey. You may need to revisit your "why" later on when you start to slide back into some of your trashy habits. The fun part (at least *for me*!) is the month on decluttering and minimalism (in chapter 2). Here you'll pare down the amount of stuff in your life so you can start your less trashy year on the right foot—a big part of reducing your waste is owning less stuff in the first place. And to avoid recluttering, you'll learn how to become a conscious consumer (in chapter 3).

Chapters 4 through 6 are where the real work begins. We'll cover fun strategies you can use to reduce your waste (think of those pesky single-use plastics) at home in the kitchen (in chapter 4), in your bathroom and cleaning cupboard (in chapter 5), and even in your closet (it's time to stop buying and tossing clothes like they're going out of style—this happens in chapter 6). Wait, you say, *they are* going out of style? Yep, we need to have a chat about that, too, because the landfills are filling up with textile waste. Don't worry, I've got your back.

In chapters 7 through 9 we explore how to be less trashy *and* still live in the real world. After all, moving to the woods and going off the grid, never to see another human, probably isn't part of the plan (though I applaud you if it is). You'll have to learn to use the word *no,* but you'll learn to do it in a way that's polite (in chapter 7). And while all these things in your life start to change in all the best ways, you'll still need to address these issues with loved ones who might be adding to your waste—or making it more difficult to reduce it (in chapter 8). This section will ensure that you do *not* become a pain in everyone's ass, but that you're still finding ways to inspire others. If your friends and family don't want to get involved and you're feeling a bit lonely, I'll also share tips for finding your own community of like-minded folks (in chapter 9).

In chapters 10 and 11 I'll cover how to budget for the lifestyle (low-waste living isn't intended to be expensive and should actually save you money—this learning happens in chapter 10). If you're really eager to explore minimalism and rein in your spending, try a no-buy month (in chapter 11). No need to go broke AF trying to be sustainable! You'll master your conscious money mindset in this section instead.

The final chapter is about bringing everything together and setting you up for long-term success, so that the changes you make extend well beyond a year of less trashy living. It's all about building lasting habits, saving the planet, and choosing progress over perfection (in chapter 12).

Ultimately, low-waste living is a balancing act, and I've come to terms with the fact that not every decision I make is going to be perfect all of the time. I've taken my fair share of criticism for not being sustainable enough, and you might, too. For any critics out there, I want to note that I've read *The Subtle Art of Not Giving a F*ck* by Mark Manson, so if you don't like my approach, I don't give a f*ck (this is said with love, I promise!). I've found my own way to ensure that the critics out there don't deter me from my mission, and I suggest you do the same. Your journey toward reducing your waste is just that—yours—and once you decide which approaches fit your lifestyle and which ones do not, there's no point in beating yourself up about what other people think. Remember that reducing your waste is not a contest. It's not about keeping up with the zero-waste Joneses on social media. It's about learning to make lower-waste choices that work in the context of your life. Internet trolls be damned!

Ready to get started on your year of becoming less trashy? Because this is your time to shine. It's your time to stand up for sustainability and have a positive impact on this planet. Individual actions matter, and you're about to prove it.

Chapter 1

Trash Talk Basics

Define your version of low-waste living, find your "why,"
and learn the eight Rs

First Things First

If there's one thing that just about everyone reading this book has in common, it's owning too much stuff. Most of us do. In fact, clearing my home of clutter was the natural first step in my journey toward reducing my waste. My fiancé and I got married in 2014. We both landed new jobs and moved back to my hometown, where he and I had met at university. Between all the generous gifts we received at our wedding, and all of the things we brought into our new home from our childhood and university years, there was quite a bit of clutter. And honestly, a lot of the junk was mine—I was even holding on to high school notebooks, textbooks, and binders bursting at the seams.

When I asked myself if I was ever going to read those textbooks or notes again, the answer was an emphatic *nope*. So the textbooks were donated and everything else went straight into the recycling bin. It felt freeing! A weight lifted that I hadn't even noticed was there. I had been lugging all of these things around with me for years, and when you keep that kind of stuff, an expectation develops in the back of your mind that you have to do something meaningful with it. But the truth was that I wouldn't ever refer back to that information, because I had better things to do with my time. In essence, decluttering those physical manifestations of my past also cleared away the mental burden of holding on to them and tending to them.

What does decluttering even have to do with being less trashy? The connection for me is about minimalism and living with less. Did you grow up hearing the "three *Rs*" adage "reduce, reuse, recycle" repeated over and over by teachers and parents? I certainly did, and decluttering and minimalism are deeply linked with that first *R:* reduce. Reducing is about contributing to a decreased demand for "stuff" by consuming less, which, in turn, helps reduce waste created by the manufacturing of products; minimalism as a lifestyle is about wanting less in the first place. Decluttering comes into the picture because it helps us start with a clean(er) slate and empowers us to let go of both physical and mental baggage. Quick word of caution: It can be easy to get carried away with the minimalist lifestyle by adopting a shiny and new minimalist aesthetic, which can lead to further consumption. It's a pitfall we'll discuss later in the book.

Whether you're into the KonMari Method popularized by Marie Kondo (asking yourself, "Does this spark joy?") or you've followed the teachings of The Minimalists, there are many effective approaches to removing extraneous stuff from your life. The link between these ideas and low-waste living is really about recognizing that we don't need a lot of stuff to live a good life. And do you want to know what the most shocking part is, when you stop to think about it? All that stuff used to be money! Our compulsion toward stuff is so substantial that we will spend an entire month on decluttering and minimalism in chapter 2.

Zero Waste vs. Low Waste

The concept for this book was born from the zero-waste lifestyle movement. That said, the *zero* in *zero waste* is an almost impossible goal to attain given our current societal norms. So what exactly does zero waste mean, anyway? The Zero Waste International Alliance defines zero waste as:

> The conservation of all resources by means of responsible production, consumption, reuse, and recovery of products, packaging, and materials without burning and with no discharges to land, water, or air that threaten the environment or human health.[1]

Blah blah blah . . . That's great, but what does that have to do with us? While this definition is useful to inform policy and legislation, it's highly technical and not relatable on a personal level. The zero-waste *lifestyle* movement, however, calls on individuals to avoid as much waste as possible. Think about yourself a bit like Neo (Keanu Reeves) dodging bullets in the movie *The Matrix*—only in this case, you're dodging trash! We live in a system that bombards us with trash, and a lot of the time it might not seem possible to escape being a trashy consumer. Yet when we slow down to inspect the things coming at us on a day-to-day basis, we find ourselves able to dodge things like single-use disposables, to avoid buying overpackaged products, and to actually send far less waste into the trash bin than we originally imagined.

When I started decluttering my home, I didn't have a lot of preconceptions about what it meant to reduce my waste, and I didn't know anything about the zero-waste subculture I would soon become a part of. These days, the term *zero waste* has become trendy, and if you've searched it on the internet, you'd probably think going zero waste means fitting an entire year's worth of trash into a small mason jar. That, of course, is followed by sharing your proud little trash jar on social media. This approach has become popularized online by some devoted advocates of zero-waste living. Sure, it's an ideal for some to strive for, but for most of us, it's often discouraging and unrealistic. From the outside, it looks like a rite of passage that separates the real tree huggers from the wannabes. Although I gave it a try, the trash jar thing never resonated with me.

Just like the weight-loss world, which has offered up extreme diets like ketogenic, Atkins, Paleo, etc., the zero-waste lifestyle movement has its share of extremes, too. I'll be the first to tell you that the tiny trash jar is one of them, and it's not really what we're aiming for in this book. What I'm here to do is to teach you how to walk before you can run—how to reduce your waste, one month at a time, to the level that works for you and your life, so that you're making changes that last. If you want to run an ultramarathon while the rest of us work up to a 5K, I will definitely be there cheering you on. But if there's anything I've learned on my zero-waste journey, it's that it can be incredibly stressful to live in modern society *and* be completely zero waste. Aside from the obvious point that it's challenging to produce basically zero trash, the guilt factor is off the charts. If I mess up and find myself stuck with a wrapper or a tag (or, even worse, *a plastic bag*), I feel like a total poseur. *Hello, impostor syndrome.* I just can't handle that kind of pressure, and I'm not here to shame you if you find a gum wrapper in your pocket one day.

I think it's time for me to make a confession: Even though I founded a lifestyle website and brand called *The Zero Waste Collective* to inspire others to reduce their waste, I'm not perfectly zero waste myself. *Low waste* is more apt, but I do sometimes use the term *zero waste* because it's the ideal goal for moving toward a circular economy (where nothing is wasted), which I wholeheartedly support.

When it comes down to it, low-waste living is about making the world a better place for future generations, and for the diverse flora and fauna that we share this beautiful planet with. So for the purposes of this book, becoming less trashy is about *the reduction of your waste as it fits within the context of your life circumstances and budget.* This is a *you do you* kind of book. Even if you make just a few lasting changes after reading this book, you will have achieved success.

The Waste Problem

In a lot of ways, the excessive waste in our lives is a result of our convenience-based lifestyles, which only really started in the 1950s, alongside the increased production of single-use plastics and other disposables. After World War II, there was a huge effort to increase production, consumption, and prosperity through a linear economic model (extract → produce → purchase → toss → repeat). To feed this hungry system, careful marketing sold us a lifestyle based on the principle of throw-that-shit-out-because-it-will-make-your-life-easier, and we bought it. We ate it all up! Literally. Fast-food chains started popping up, and more people were driving around in their cars to enjoy this groundbreaking convenience and newfound freedom from chores like dishes. Then *BAM!* The one-and-done lifestyle became mainstream. (I mean, I wasn't around for the transition, but from what I've read, it all happened pretty quickly.) And here we are now, in the twenty-first century, and it's normal to just throw stuff out. Yep, being trashy is the norm.

Planned obsolescence is no secret, either. What does that fancy-pants term mean? Well, it means that a lot of the stuff we buy—from appliances to vehicles to household supplies and electronics—started being designed to have a shorter life span. If these products didn't last as long, we would have to buy more crap to replace them more often. Just think about that

Depression-era stove your grandparents had in their kitchen. It was probably built like a tank, made to withstand Armageddon and last *forever*. Most stoves just aren't built that way anymore; they are designed to have a shorter life span and to be replaced often. Unfortunately, this approach became more popular in the early to mid-1900s to increase sales and profits and still persists today. Using weaker parts to build a product so it doesn't last as long is one of the most obvious forms of planned obsolescence, but there are many other ways companies conspire to keep us buying and tossing:

Function: Products are designed to break. Instead of being designed to outlive us, stuff tends to fall apart or stop working quickly and easily, so that businesses can sell more replacements and reap more profits. There goes that paycheck. . . .

Style: Trends change so quickly that last season's (or last month's) wardrobe is no longer in style and we frequently feel the need to clear our closets so we can fill them with the newest trends. Does this happen in *your* closet? If so, you're not alone.

Software upgrades: Need a new computer or mobile phone every couple of years because your hardware can't keep up with software updates? That's planned obsolescence too, and it's producing as much as fifty million tons of electronic waste worldwide.[2]

If you've ever wondered why you had to replace a toaster that's only a few months old, you can blame planned obsolescence. That toaster was destined for the landfill before you even bought it. That's because it was designed to break quickly, prompting you to buy a new one again to replace it. While some of us may take the extra time to fix something ourselves, or have someone else fix it, some things are also designed to be difficult to repair. When products are challenging to fix, it leaves us (the consumers) with no choice but to make a new purchase. To break this cycle and reduce the waste it creates, it's important to buy things that are built to last.

We didn't personally create the convenience-based and wasteful world we live in today. Instead, we were born into it and conditioned to fulfill this extract → produce → purchase → toss → repeat sequence. It's unfortunate that you and I are left to clean up a mess we didn't make, but there's not much we can do to change the past. What can we control? Ourselves. We get to decide (for the most part) how we want to live and how our purchases can contribute to a less trashy future.

While it may be tougher these days to buy things that last a lifetime (so much stuff out there is crap), it is still possible. With the internet making information easier to access, we have the opportunity to do a bit more research on the products we buy, get to know the companies we're going to purchase from, understand their warranties and repair policies, and read product reviews. With the tools provided in this book (especially in chapter 3, "Conscious Consumption"), you'll become a stealthy shopper and smell planned obsolescence miles away. No more getting duped, my friend.

Let's go over the steps coming up in this chapter that will form the foundation of your low-waste journey: conduct a trash audit; relearn the three *R*s: reduce, reuse, and recycle; add a few more *R*s: rot, refuse, repair, repurpose, and rethink; and find your "why." The new, less trashy you starts now!

Conduct a Trash Audit

When you start on your path to low-waste living, it's good to know what type of waste you're creating in the first place so you actually know what changes will most effectively reduce your waste—and even eliminate some types of waste altogether. After all, you don't know what you don't know . . . until you learn more! To complete this trash audit, you're going to collect all of your trash, recyclables, and compostable items over the course of one month and analyze what types of waste you're creating at home. One month is long enough to get a sense of the common types of trash you toss out, whereas if you did this activity for only a week, you'd probably miss quite a few items. The purpose of this activity is to help you determine areas where you might be able to reduce your trash.

But first it's important to understand your municipality's waste-management system. While this probably sounds like a boring exercise, you might discover that you're creating more waste than you thought, once you familiarize yourself with what actually can and cannot go into your recycling and compost bins (if you use them in your area). Residual peanut butter in a jar, for example, has the power to prevent cardboard from getting recycled if it gets greasy from the peanut butter. That's why it's important to clean out containers before popping them into your recycling bin. It's also essential to know what is considered hazardous waste and to learn whether you're supposed to drop certain things off at the landfill in person or if there are special pickups available.

You may be able to find all of this waste-sorting information on your city's website or on a mobile app created by your municipality. If you can't find info using these resources, try calling city hall for the information instead. You may be able to get a hard copy of a waste categorization list (or print it from the website) to put on the fridge. Keeping these details visible will make it easier for everyone in your household to sort waste.

To complete your trash audit, first grab a notebook or open a new spreadsheet. Create a chart and list applicable waste-separation categories across the top. Make sure the ones you choose align with how your city separates waste. My chart headings looked something like this:

Landfill	Recycling	Organics	Hazardous Waste

Once you've developed your chart, be it in a notepad, on your phone, or on a computer, let the following steps guide you through the process!

1. Establish a set period of time (e.g., one month) to audit your trash (or your audit can be ongoing).
2. Involve all family members or roommates in the discussion so you can get everyone contributing ideas and buying in to the solutions.
3. As you dispose of things, list each item under the appropriate heading, or grab gloves and sort through waste that's already in your bag or bin, adding each item to the list.
4. Review your findings and tally up all of the results to better understand how much of your waste is going in each stream.
5. Reflect on these questions:

   - Are there any patterns emerging in terms of the types of waste you tend to throw out the most?
   - Now that you know your recycling system, are there ways you can reduce your landfill waste and recycle better?
   - What goals do you want to achieve over the next year to become less trashy?

- Are there ways you can reduce the materials you place in your recycling bin?

6. Keep the results of your trash audit handy as you go through the rest of this book, as it will help you to know which areas you need or want to focus on the most.
7. Conduct another trash audit at the end of your Don't Be Trashy Challenge, and compare it to where you started to see your success.

Trash audits can be ongoing and even recurring. For example, you could choose to do a home trash audit annually, or even semiannually, which would help you determine if you're successfully reducing your waste. Knowing your current waste-production habits from your trash audit is an important part of moving toward a less trashy lifestyle. If getting Uber Eats is your weakness, for example, you'll be able to see that clearly in your chart—because of all your documented takeout containers—and then you can use that as one area to focus on to reduce your waste.

The remainder of this chapter and the rest of this book will help you create your own road map to become less trashy; this trash audit simply lays the foundation of knowledge that you can use to make the most of your lifestyle changes and see the positive results from your Don't Be Trashy Challenge. Hopefully by the time the year is up, you'll be sending way less trash to the landfill (and less into recycling, too).

Relearn the Three *Rs*: Reduce, Reuse, and Recycle

I was in grade two or three when my city started composting municipal waste. When the city introduced the program, everything was sorted in bags.

There was a clear bag for landfill waste, a clear blue bag for recycling, and a clear green bag for compost. These bags were all transparent so city staff could decipher whether or not you had sorted your waste properly, and if you hadn't (tsk, tsk), they'd leave it with a note explaining how to sort it properly. Having to re-sort trash because you didn't do it properly the first time totally sucked, but it was a great way to learn.

When this transition happened, city staff dressed up in big teddy bear costumes representing the three waste streams and came to my school to teach us kiddos how to sort our waste properly, and to educate us on the three *R*s: reduce, reuse, and recycle. It was cheesy, but it was so effective I still remember it decades later.

It turns out they were right to put on a big show about the three *R*s, because they're still so important today. Chances are, even if you had a formative experience of the three *R*s like I did, you've probably lost touch with actually putting them into practice on a regular basis. So let's go back to basics and revisit them:

1. **Reduce:** This one is first for a reason. If you reduce what you consume, you'll produce less trash. Full stop. It's a super simple concept but not always easy to do. Reducing is about both paring down what you have and need and reducing your consumption moving forward. For example, if you declutter and clear out your wardrobe, you've pared down. However, to maintain your pared-down wardrobe, you'll also need to reduce the amount of clothing you buy in the future; otherwise you'll quickly reclutter your closet. Reducing by consuming less is better for the planet, because you'll reduce your reliance on the resources necessary to produce what you consume. The more you reduce, the less you'll have to reuse and recycle.

2. **Reuse:** Ditch single-use disposables of all kinds and reuse instead! Choosing reusables instead of disposables is often what comes to mind for those who are familiar with the zero-waste movement, but it's not always easy to do when you're eating on the run or decide to make an impromptu grocery stop after work—unless you prepare in advance by putting together your own zero-waste kit. What's in a kit? Well, it's up to you. It can include any reusables you find you need for life on the go. Here's a list of things I keep in my tote bag just in case I need them:

- Water bottle
- Coffee cup (admittedly, I only drink tea!)
- Straw
- Napkin
- Handkerchief
- Eating utensils
- Food containers / jars
- Shopping bag(s)
- Produce bag(s)

Other ways to reuse include using what you have, or what others have, instead of buying brand-new. It's about reusing what has already been produced, made, and consumed. These reusing options include shopping secondhand; borrowing from and lending to friends and neighbors; swapping stuff with family and friends; renting things like tools (like a drill set) and specialty equipment (like a carpet cleaner); and if necessary, buying products that can be used again and again, that have great longevity and repairability. Reusing helps curb consumption, reduces our packaging waste, and keeps us from buying unnecessary crap and from relying on single-use disposables like coffee cups and straws.

3. **Recycle:** "But I recycle!" and "I recycle, too!" are common refrains I hear when I share that I live a low-waste lifestyle. Yes, recycling is important, but it's not the answer or solution to our waste problem. Unfortunately, most recycling rates worldwide are so low that a lot of the materials you put in your recycling bin still end up in landfills, get incinerated, or make their way into the natural environment. If you want to reduce your waste, it's best to avoid as much material as possible that will end up in either your trash bin or your recycling bin. However, if you're going to toss materials into your recycling bin, it may be best to find products that are most commonly recycled where you live. Call your local waste-management facility to find out, but here's a good tip: The simpler the material, the simpler it will be to recycle (think paper, aluminum, or glass). Many packaged products these days are made from a combination of materials, including a combination of paper, plastic, and metal. When all of those different materials are fused together, they become much more difficult, expensive, and energy intensive to recycle.

Remembering the three *R*s will probably be the simplest way for anyone to reduce their waste. Most people reading this book will already be familiar with them, and now it's just time to put them into practice once again. That said, if you're ready for more (I know you are because you're reading this book!), let's dive a little deeper into more *R*s that can guide your life to be less trashy.

Add These Five *Rs*: Rot, Refuse, Repair, Repurpose, and Rethink

Why stop at the three *Rs* when we can add a few more that are just as reliable when it comes to reducing waste? While these *Rs* are not part of the core group that we all learned growing up, they're still helpful in situations that don't necessarily involve the first three *Rs*.

4. **Rot:** Rot is another way to think about compost. Composting is an incredibly important part of reducing landfill waste, because organic material (like food waste and tissues) creates toxic methane gas (a greenhouse gas contributing to climate change) when it rots in a landfill with other garbage. Food waste should never have to go in a landfill when it can easily be recycled into compost, which is rich in nutrients and can be reused as a fertilizer for agriculture or home gardening. While some municipalities offer composting as an option, you can also do your own composting at home in your backyard, on your apartment balcony, or even under your sink with the right kit. We'll chat more about the available composting options in chapter 4. For now, keep this important step in mind if you're currently throwing your organic matter in the garbage. The best part? You'll *significantly* reduce your waste and environmental impact if you make composting part of your routine.

5. **Refuse:** Refusing is essentially saying no to unnecessary and unwanted crap. It's such an important step that the entirety of chapter 7 is devoted to it! It's well aligned with *reduce,* from the three *Rs,* but it's almost like the precursor to it; in order to reduce, we must refuse. Say no to things like junk mail, subscriptions, single-

use plastics, freebies, and samples. I have a "No Junk Mail" sign above my mailbox, we don't get the free local newspaper delivered anymore because we don't read it, and I don't bother to take the freebie bags that are often given out at conferences. I also don't use the small toiletries at hotels—I bring my own, and you can, too!

6. **Repair:** With planned obsolescence, it's more common these days to toss something out that's broken instead of repairing it. That's often because it's cheaper to buy a new replacement than it is to repair what you have, because the replacement parts and/or labor are so expensive. Instead, choose to buy things that are repairable, and repair the things you already own if possible, in order to reduce your consumption and disposal of products. In my community, the local tool library has repair cafés, where handy people come out and volunteer to fix broken stuff *for free*! Find out if a similar resource is available where you live; if it's not, maybe you can set it up. To reduce our waste, trashing broken stuff should be a last resort. Repair first!

7. **Repurpose:** Repurposing an item that no longer serves its original purpose is a great way to keep it out of the trash. This is basically upcycling, which entails the use of waste materials to create something new. From collecting and using wine corks to make wall art to saving old wood window frames to make a rustic mirror, there are many creative opportunities to make old things new and loved once again. Next time you're not sure what to do with an old something or other, see if you can find a way to repurpose it and keep it out of the trash! Keep in mind, though, this isn't intended to be a reason to start stashing everything away. If you can't find a use for something and won't repurpose it yourself, give it away for free so someone else can enjoy it!

8. **Rethink:** All of the *R*s above are about finding new ways to reduce our waste. They're about looking at our stuff in new ways and considering the full life cycle (extraction, production, waste) of the products we buy to examine the bigger picture of our consumption habits that eventually result in waste. When we start to rethink our lifestyle, change our habits, and look at our lives through this new lens of waste, we can more easily reduce the waste we produce. Rethinking on an ongoing basis empowers us to employ all of the other *R*s to successfully be less trashy.

Find Your "Why"

I wish I could flick a switch or snap my fingers and all of a sudden, everyone in the world would care about living more sustainably. Governments would build policies and legislation to ensure that companies and citizens operated in ways that respected people and the planet. They would put systems in place to make it easier for citizens to live more eco-friendly lives, and less trashy living would suddenly become the norm!

Yet I can't make anyone care with the flick of a switch—not even you. Sure, I can tell you about all of the ways that reducing your waste will reduce your climate impact, how it will help conserve resources and minimize pollution, or how it can help foster a sense of community, but if *you* don't really, *really* want to make a change in your life, you probably won't—at least not a long-term change.

I don't want to tell anyone how to feel or how to live. It's up to them, and it'll be up to you, to decide what's important and what changes you want to make. I can show you the greener path I've found through my own exploration, but it's up to you to choose how you want to move forward.

Take a moment to dig deep and reflect on why you picked up this book.

I'm assuming you want to live more sustainably, right? How come? What's your motivation? What do you want to change? Is it to do your part to slow the rate of climate change or perhaps reduce the plastic waste that's compromising marine life around the world? Or maybe part of your reason for wanting to make a change is a little bit self-focused! That's okay, too. If what you really want is to save the world while also saving yourself some money (prepackaged food can cost a lot!) or lose some weight (prepackaged food is often less healthy than fresh, unprocessed stuff!), use those as your reasons.

Whatever reason(s) you settle on for wanting to make a change, make sure you care about them a lot. They need to be significant enough to keep you motivated to start and to continue making small, incremental changes each month. So take a minute or two to ponder. My aha moment happened in Bali. What's yours?

Even when you have a really good reason for making a change, there can still be moments when you want to return to some of the habits of your old lifestyle. Even if they're not necessarily easier, they might feel easier because you've been doing the same things for so long that they are automatic. If you ever find yourself losing motivation, here are some tips to keep you going on this journey:

Educate yourself: Always keep learning! No one on this planet knows absolutely everything there is to know, so you know there's always something more to learn. Keep learning about the issues you're most passionate about!

Remind yourself about what's motivating you: Once you've found your "why" by determining the main issues that you're passionate about, revisit them when you feel like you're about to fall off the wagon (you probably will at times). Remember why you took on this project in the first place. I like to rewatch the documentary *A Plastic Ocean* to remind myself why I avoid single-use plastics as often as possible.

Create a personal mantra: For a quick daily reminder of what you're doing and why, develop a mantra to repeat daily, like "Less is more" or "Do this for the planet" or even "Don't be trashy." My personal mantra when it comes to consumption more specifically is "Buy less, buy better, and skip single-use!" It's a bit simplistic and doesn't cover all of the ways in which I aim to reduce my waste, but it's a quick reminder to fall back on when I start to think about shopping or eating out.

Reflection and Checklist

This chapter was about going back to basics. Revisiting the three *R*s and doing a trash audit sets the foundation for what's to come in the rest of the book. The following chapter is all about paring down and training yourself to consume less. Before we hop to chapter 2, take the time to follow the steps outlined in this chapter. Here's a recap:

- Do a trash audit and make a list of goals to reduce your waste over the coming year
- Reacquaint yourself with the 3 *R*s: reduce, reuse, and recycle
- Create your own zero-waste kit for life on the go
- Add a few more *R*s: rot, refuse, repair, repurpose, and rethink
- Find your motivation—your "why"—for becoming less trashy
- Create a lifestyle mantra or mission statement to remind yourself why you want to make some changes to your everyday habits

Chapter 2

Decluttering and Minimalism

Let that shit go, and don't replace it with more shit

My Closet Is Too Small

I remember this moment all too vividly. My husband and I had just moved into our first home shortly after getting married, and it was hard not to stumble over boxes and furniture that had yet to find their place. Built in the early 1900s, our modest two-story redbrick house came from an era long before the McMansions that we see popping up today. With a modest house come modest closets. Wayyyyyy too modest. You *could* call our master bedroom closet a walk-in closet—two of me could fit crouched inside—so long as it was empty. I quickly filled my dresser drawers and this "walk-in" closet, but I was left with a mountain of clothes piled up on the bed, sadly awaiting their unknown fate. On the surface, getting rid of the excess seemed easy. The reality was that it would be emotionally demanding to sort and let go.

The truth is, I hadn't seen most of these clothes in what felt like an eternity. Having been a student for many years and being newish to the working world at the time, I had moved so frequently that I kept a storage unit (*gasp!*

I know). This was to avoid lugging all my crap from city to city and apartment to apartment.

It might sound strange for a twentysomething to have a storage unit, but I'm not the only one. Most people simply use their parents' houses as an unofficial storage unit for the snowboard they bought that one year they joined the ski club in high school, their electric keyboard from band practice, and maybe even a lifetime's worth of birthday cards that feel too special to throw out—all the stuff they might want one day but don't need right now (and in many cases haven't needed or even looked at for years). If you're like my husband, maybe you never fully cleared out your childhood bedroom and didn't bother to sort the trophies and medals from high school sporting events, among other things like clothes and band posters.

As I confronted the Mount Everest piled on my bed, I felt the clothes staring back at me with disdain. I had let them down; they were unloved and unworn. I distinctly remember pulling my hot-pink tube top off the top of the pile; its heyday was six years earlier, and I'd known for a while that we didn't have a future together. So why was I hanging on? Probably because we shared a past that was fun, adventurous, and even a bit wild at times! That pink tube top had been with me through my university years and traveled in my suitcase to Poland for my semester abroad. I vividly recall going out to underground nightclubs in Kraków's Old Town. With their centuries-old stone and brick walls, entering these nightclubs felt like stepping back in time, until you registered the modern music, lighting, and all of the vodka flowing from the bar. It seemed to me that discarding my pink tube top was kind of like discarding fond memories and past experiences. In reality, though, these memories weren't going to disappear with the top; they would live on in my mind and in the photos I had. With that realization, I decided it was time to break up and move on.

I'm not sure what I would have done if I'd had a larger closet back then.

I might have shoved my unworn clothes somewhere in the back. That would have been the easy way out: to avoid decluttering altogether instead of confronting the emotional baggage that often comes with letting go of physical possessions. In my case, I didn't have any choice but to pare down. I was forced to declutter and to deal with my emotions head-on because these excess clothes had nowhere to live. As a self-identified tidy person, I wasn't going to let them sit on my bed. That pile had to go, and that physical process of decluttering gave me the mental space I needed to fully embrace the next chapter of my life in our new home as newlyweds.

Sorting through my unworn clothes and accessories was a blessing in disguise. It made me reevaluate what I needed from my wardrobe. I had a real lightbulb moment when I came across a pair of unworn, knee-high cognac boots with a three-inch heel. While I was brought up on *Sex and the City* and clearly had dreams of building Carrie Bradshaw's shoe closet, I live in a small, suburban city called Guelph, about forty-five minutes west of Toronto, Canada. This is not New York City, people! These boots were for my *fantasy self,* but in real life I wore Birkenstocks all summer and thermal, waterproof boots all winter—and these remain my closet essentials today. Those cognac boots might be right for you, and if they are, then you should keep them, but unfortunately, they didn't have a space in my *real life.* Clearing out my clothes and shoes really helped me better recognize who I was in that moment and how I wanted to show up in this world. This is really important, so we'll talk more about it in chapter 6, "Outfit Repeater."

While my unintended hoarding was inconvenient, a bigger issue became obvious to me years later: trying to keep up with the Joneses. We live in a consumer-driven society, and we're constantly bombarded with the message to buy more than we need. The messages are typically subliminal, of course; marketing campaigns often approach us with our fears and insecurities in mind. Their suggestions to "buy this to look more beautiful" or "buy that to look rich" or "buy this to lose weight" make it easy to buy in excess. We feel like we're solving our perceived personal problems by purchasing products dedicated to eliminating them. All of these superfluous purchases drain our bank accounts, fill our homes with crap, and leave us feeling empty because they don't make us happy. When I look back, I see that my mountainous pile of clothes was perhaps a reflection of an attempt to keep up outer appearances, all the while putting a Band-Aid on my true insecurity of wanting to fit in with those around me. So to hell with the Joneses! No matter how hard we try, there will always be people who have bigger houses, faster cars, and trendier shoes. They'll probably look more beautiful and smell better, too.

In this chapter, we'll get to the root cause of our clutter and consumption. Are you ready to peel back the layers of your consumption patterns by cleaning out your clutter? When you sort through the stuff you own, you can more clearly see what you regularly consume. It's this process that will empower you to create better shopping habits and, of course, waste less while you're at it! Don't worry if this clean-out seems like a lot of work; just take it at your own pace. The process I outline below is one that I revisit frequently. As our lives change, so do our consumer needs, which is why it's important to check in with yourself now and again if you plan to keep your clutter at bay and keep your shopping habits in check.

One of the hidden bonuses you'll discover as you begin to declutter is that an uncluttered home builds the foundation for an uncluttered mind. Going through the practice of letting go can be an emotional expedition and a physically tedious activity, but the reward is more simplicity in your life. Keep the end in mind if you start to feel tired. You'll be working toward unloading the emotional layers and physical clutter and swapping them out for less stress and a lighter home; adopting a minimalist mindset and consuming less, which saves money, resources, and waste; and ultimately, understanding yourself so well that you become confident from the inside out, not the outside in.

And what's this about minimalism? Well, it doesn't mean you have to strip your home bare and get rid of all your clothes and join a nudist colony (unless you want to, of course!). It's about choosing to live with less. And while *owning* less is great, *desiring* less in the first place is even better. Because here's the thing: Just because you declutter doesn't mean you won't revert to your old ways. You could end up quickly filling all the spaces you've cleared with new things! Before you jump on the bandwagon and go shopping again, consider taking a breath and truly evaluating your wants and desires. While minimalism may sound like you'll be going without, it can help you focus on the more important parts of your life that bring you the most joy, like spending time with loved ones or pursuing hobbies. A minimalist lifestyle doesn't need to be extreme; you can make it what you want. No one gets to be the judge except you.

Let's take a deep breath and get a sense of what lies ahead: First you'll draft your Minimalist Mission Statement. Then we'll follow that up with the process of letting your crap go, and how to do it responsibly. Through all of this, we'll uncover the connection between minimalism and waste. And finally—and this is probably the most important step—I'm going to help you develop your own tactics to avoid recluttering your space. Let's do this!

Minimalism and Letting Your Crap Go

In addition to dealing with my limited closet space, having to sort all the other junk from my storage unit was what eventually led me to discover the minimalist lifestyle movement. One night, feeling overwhelmed by my excess, I scrolled through countless articles on my phone looking for creative ways to store my stuff. My mindset was all about better storage and organization. How could we maximize our small space? I didn't want any area in our home to become a dumping ground—I'm Miss Tidy, as you may recall.

There is, however, a big difference between being tidy and being mini-malist. You can be tidy and still have a lot of crap; it might simply be well hidden or neatly organized. And who knows, maybe there are messy mini-malists? But at least their messes will be quick and easy to tidy, because they don't have much stuff! On my late night with Google, I went down the minimalist lifestyle rabbit hole on the internet and discovered that *owning less can be easier than organizing more*—thanks to Joshua Becker's blog, Becoming Minimalist, for that revelation! I didn't need to go to the Con-tainer Store or Homesense to stock up on pretty baskets and storage bins of all sizes and colors with fancy labels. Rather than spend more money on these so-called storage solutions, my new plan was to lighten the load. This eye-opener, I'm pretty sure, saved me hundreds of dollars and many hours, or even days, of organizing. Therein lies the big, juicy secret. It's not about creative storage solutions; it's about having less to store in the first place! And minimalism is not just about getting rid of our personal things to live with fewer physical possessions. Our availability to enjoy life can be increased by reducing the number of hours we spend finding, organizing, cleaning, and maintaining our stuff.

At this point, you might be nodding excitedly with agreement but also wondering how to get started! Before you jump headfirst into decluttering, it's important to explore why you want to undertake this activity in the first place. After that, I'll provide a few simple strategies below for decluttering your home, followed by tips about what to declutter. If you were already a minimalist before reading this chapter, it's still worthwhile to revisit any stuff you may have accumulated since the last time you decluttered. Chances are, you'll find some things—perhaps many things—that are going unused and unloved in your home, even if you don't think of them as clutter.

Reasons to Declutter

You may find that minimizing your possessions does not feel like a priority for you. It's something I love to do, but as soon as I turn to my husband to join me, he runs the other way. Maybe that is your gut reaction, too. While it may not be your cup of tea to tidy up, nearly everyone can enjoy the benefits of decluttering. Take my husband as an example. While he hasn't openly admitted this to me, I've noticed he gets excited about finding buyers for his unwanted things online using Kijiji, an online marketplace for mostly secondhand goods that's popular in Canada. Whether he makes twenty bucks selling an old drill or sixty for his ancient Xbox, I've noticed that he lights up when he earns some cash on the side for stuff he no longer uses. To him, tidying up means making money!

Unlike my husband, I get excited by the prospect of more space and less mess in my home; it's emotionally freeing and leaves me with fewer chores. What's important here is to find what motivates *you* deep down. A good question to ask yourself as you embark on this personal assessment is *Why do I own all this stuff in the first place?* Go ahead, give it some thought. This might help you think about why this exercise is worth your while. Once you figure out what you're going to get out of decluttering, try developing a Minimalist Mission Statement. The mission statement should outline your main motivation(s) for decluttering and owning less. Here are the two key elements to include in your Minimalist Mission Statement:

1. Your purpose (the main reason or reasons you're doing this)
2. Your desired outcome (how you envision your home and your life)

Take a moment to write down all of your ideas. You can also do this in your head if you're not the journaling type. Do a complete brain dump of why you want to declutter your home and why you want to live with less, and follow that up with how you envision your home and your life as a result. Then pull the most important ideas into your mission statement and write out the final version. Here are some ideas to inspire your list of reasons why you may want to declutter and live with less:

- To create a calmer living space that feels and looks peaceful
- To have less to clean and dust
- To have less to organize and store
- To let go of the guilt associated with certain items (e.g., unwanted gifts)
- To inspire creativity in your kids
- To have a calmer household, if your kids are happier with fewer toys
- To give unused items in your home the new life they deserve, with people who will use them
- To make some extra cash by selling unwanted things
- To let go of the burden of maintenance
- To learn your consumption habits
- To inspire yourself to shop less
- To have more space (both mental and physical)
- To focus on more important things in life that bring you joy (keeping in mind what those things are for you, which could be family time, hobbies, etc.)
- To love yourself from the inside out
- To dig deeper into who you are and who you want to become in this life
- To give back to your community with your donations, time, energy, money, etc.

- To let go of one life chapter and move on to the next
- To downsize your home

Once you've written out your mission statement, you can return to it as you go through the decluttering process. Keep in mind that it can be updated as your needs change, too! If your excitement for decluttering wears thin or you run out of steam, then revisit your Minimalist Mission Statement to refresh your journey to living with less!

How to Declutter

My personal rule of thumb is to keep only what you use, what you love, and what you believe to be beautiful in this season of life. The "season of life" is a key component here that is not to be confused with the seasons of the year or the seasonal usefulness of the things you own, like a winter coat, which you'll need to keep for the following winter. The "seasons" or phases of life are defined by the transitions we go through over the course of our lives, which often result in our having different needs in each phase. For example, parenting is one season of life, and being an empty nester is a different season of life. Consider what season of life you are in when you declutter!

When you're decluttering, it can be easy to rationalize keeping more things than you need to. You might think, *Well, I will probably get around to using that one day,* or, *I used to adore this, so that's why I still have it,* or *I thought this was gorgeous when I bought it, so I should keep it.* When you choose to weed out things that do not serve you in *this* season of your life, you're letting go of things that are not useful in the present—even if they were useful in the past or there's a small chance they might be useful in the future. Items that spend the majority of their time hiding in the back of a closet or up on a shelf might be better off in the hands of people who could use and enjoy them *right now*. So if your kids have outgrown their clothes and you don't plan to have any more littles, consider letting them go. Decluttering these things can be easier said than done, but I have some suggestions that will help you through the process.

A word of caution before you get started: Focus on your own stuff first! If you live alone, you won't need to worry about this. But if you live with family or friends, then stick to your own belongings, no matter how tempting it may be to quietly toss your partner's ugliest shirt or holey boxers. If you declutter something that isn't yours, or something that's considered a shared item, like a board game, without discussing it first, you could be asking for trouble—and trouble may tempt you to abandon the whole endeavor.

You can also use the time spent going through your own things to drum up household excitement for future decluttering projects. Tell everyone you share a home with what you're up to and invite them along to participate! One way to lure them into your endeavors is to share with them all of the benefits of decluttering, particularly the ones you know would trigger their motivation (like my husband with the bonus cash). If they're still not interested, don't worry. They may see the fruits of your labor and choose to get on board later.

Now let's get you going on your decluttering journey! Start by thinking about or writing down a list of the things that you love the most—the items that you use every day. A few of my most prized items include my international Starbucks mug collection as a reminder of the cities and countries I've visited; sentimental jewelry that I enjoy wearing often; and the books I reread the most. This is not an exhaustive list of all my prized possessions, but hopefully it'll give you an idea of what to include in your own list of favorite things. These are the things you love so much that you know you won't be decluttering them! Just be careful not to put *everything* on that list!

Then move on to the items you know are collecting dust. These are the things that are sitting at the back of your closet and kitchen cabinets or under your bathroom sink, and perhaps in the depths of your garage or shed. Put these items on a separate list. Although it's been some time since I did a big purge, here are some of my belongings that sat in my home unused for too long: a panini press (i.e., sandwich grill) that I thought would make cooking easier but collected dust instead; Rollerblades that I used for years, until I didn't, and they took up precious space in the shoe closet; and an iPod Touch that was replaced by an iPhone and sat unused in a desk drawer. Thankfully I sold each of those items with ease.

It probably won't take long for you to come up with a list of things in your home that haven't been used in years. Consider those pants that haven't fit in over five years, the stunning costume jewelry you bought for your high school graduation, or the electric wine opener taking up a quarter of your utensil drawer. Clearing out your clutter is one way to make sure that what you keep reflects your current lifestyle needs and interests and not those of your past, or even your aspirational future. Sometimes, even if you hold a lot of emotion and/or nostalgia toward physical items, it can still be time to move on; this is especially true when letting go becomes mentally freeing, like donating the pants that no longer fit. Even though I enjoyed my Barbies *so much* as a little girl, there came a point when I knew it was time to let another child find happiness in my Barbie collection. While that might seem like a simplistic example, it was actually quite symbolic of my transitioning from a young child to a big kid. Decluttering can be a physical manifestation of our life changes. A good question to keep coming back to is *Does this item reflect what I use, love, and/or believe to be beautiful in this season of my life?*

Now it's time to dig in and start getting rid of stuff! I find that the most effective approach to decluttering is to bring the right category (e.g., paperwork) of stuff to the right location (e.g., home office). The idea behind choosing a category to tackle instead of just a location is that our stuff tends to scatter to places it doesn't belong. Let's take paperwork, for example. In an ideal world paperwork would reside in a designated location, like a filing cabinet. In reality, it tends to live at the front entryway, in the kitchen, on the dining room table, *and* in the office. When dealing with a category of stuff that often creeps into areas outside of its home, tackling a specific location might not get the job done. It's best to collect all of the paperwork and bring it to its designated location (like your home office) and declutter there. If all of your paperwork is already in the right area, but it's a disaster, then work in that location to sort and declutter. *Note: Lists of where and what to declutter are outlined shortly!*

When it comes to decluttering, it's most effective to label designated piles, such as keep, sell, donate, recycle, and trash. You can use bags and/or boxes or make clear piles on the floor, as long as you know what each pile represents. Using actual labels may be helpful, especially if you have helpers and you don't want your "keep" pile to end up at the donation center! Don't be trashy, and make your trash pile a last resort whenever possible (to avoid waste, of course). Once your piles are clearly defined, it's time to do the work of sorting your stuff. As you go through your things, focus on one object at a time. You get to create the standards you'll use to determine whether or not an item gets to take up precious real estate in your home. Often you will have a strong intuition about which pile an object should belong in, but some items will be more difficult to sort. When you hit a roadblock, try asking yourself:

- Has this expired? *Yes? Throw it out! No? Are you going to use it before it does? If not, pick a pile and toss it in.*

- Is it in working condition? *No? Can it easily be repaired? Do you want to repair it? Will you actually repair it? If yes, take steps to do it before the end of the week. If not, you know where it belongs.*
- When was the last time I used this? How often will I use this? *Be realistic about when you will actually use an item again. If you're not sure, give yourself a deadline. If it's not used by then, it's time to pick a pile and let it go!*
- Do I have multiples of this item? Do I need multiples? Could owning just one do the job? *If you have multiples of an item, consider this: Can you actually use more than one at any given time? If so, would you realistically need to use more than one at any given time? If no, then let it go!*
- Is it worth my time to clean and maintain this item in the long run? *Are you really going to take the time to clean and look after it? If yes, then keep it. If no, then choose its proper fate!*
- Do I really want to store this? Should it really be taking up space in my home? *Does it serve your current season of life? If so, keep it; if not, pick a pile to let it go.*

The list above is effective if you're not having trouble letting go. If you're having a particularly tough time deciding whether to get rid of something, here is an additional list of questions to consider:

- Do I feel guilty for getting rid of this? If so, why?
- Am I keeping this because it was a gift?
- Am I keeping this because someone else thinks it's worth keeping?
- Am I keeping this because I paid a lot of money for it?
- Do I have this because it's trendy?
- Am I keeping this to show off to other people?

- Do I want to keep this because I think I'll need it someday? When is someday?
- Do I keep these clothes because I want to fit in them one day?
- Am I keeping this because it represents a life that I'd like to live (think back to my shoe closet worthy of Carrie Bradshaw), but that doesn't reflect my reality?
- Am I keeping this because of nostalgia and therefore holding on to a past chapter of my life that is no longer relevant?
- Am I keeping this because it represents the life of someone who has passed away?
- Do I keep this simply because I inherited it?

Some of these questions may be tough to get through. Your responses to them should highlight the true reasons you're struggling to let go. At the end of the day, stuff is stuff. We are the ones attaching our emotions and memories to physical objects. Just because we let go of an object does not mean we're letting go of the emotions and memories associated with it. When I'm missing the days of wearing my hot-pink tube top that I decluttered, I look at photos that let me stroll down memory lane. Similarly, as you let go of things from your home, consider taking photos of those really special items so you can keep visual mementos to spark specific memories.

If you plan to declutter things that are important to your family members (perhaps even those you don't live with), be sure to ask them if they'd like the item before you get rid of it. It's not worth upsetting loved ones who might want to keep a family heirloom you're not fussed about.

A quick note about guilt. It can be tough not to feel guilty about giving away unwanted gifts. A gift can be a person's way of showing love and gratitude, so it often feels like receiving a gift is receiving their love and gratitude. But what you choose to do with that gift afterward is completely up to you.

Sure, we can debate this, because you might say something like "Well, if my granny comes over for tea and doesn't see the special teapot she gave me for my birthday, she'll be very upset!" Consider it this way: Is your granny visiting you just to check on the teapot? Or is she visiting you because she cares about you and loves you and wants to spend quality time with you? It might be nice for her to see the teapot in action, but at the end of the day, it's quality time that's most important. To avoid getting unwanted gifts in the first place, you'll master the subtle art of refusal in chapter 7.

Now that you're equipped with the tools to declutter, go for it! Pick a category of stuff to work through, choose a location to work in, and sort your things into your designated piles. Check out the following sections for *what* and *where* to declutter. You are now officially on the road to simplicity!

What to Declutter

What should you consider decluttering? *Everything.* That doesn't mean you'll get rid of everything; it just means you'll assess all of the things you own to make sure each item meets your new standards to stay in your home (think back to your Minimalist Mission Statement). To simplify the process for you, and to help you home in on the categories you might want to work through, here's a list to keep you on track:

- Baby clothes and accessories
- Books and magazines
- CDs / DVDs / video games
- Cleaning supplies (see chapter 5)
- Clothes (see chapter 6)
- Coats and jackets
- Collections/collectibles
- Construction materials and scraps
- Craft supplies

- Electronics
- Entertainment and party supplies
- Extra furniture
- Food (see chapter 4)
- Games
- Gardening supplies
- Hobby and sports equipment
- Holiday and seasonal decor
- Home decor
- Jewelry
- Kids' clothes and accessories
- Kitchenware and small appliances
- Knickknacks
- Laundry accessories
- Linens and pillows

- Medications
- Office supplies
- Paperwork and mail
- Pet supplies
- Photos and memorabilia
- Plants and pots
- Pool stuff
- Purses/backpacks/briefcases
- School supplies and books
- Shoes
- Toiletries and makeup (see chapter 5)
- Tools and equipment
- Toys
- Unfinished projects
- Unwanted gifts
- Vacation accessories
- Vehicle accessories

There's a lot on this list. Humans seem to require a lot of stuff! Let's walk through one category together. School supplies and books were a source of clutter for me that I would keep from year to year. So much so that by the time I graduated from university, I had accumulated notebooks, textbooks, pens, and pencils over a period of two-plus decades! That's a lot of junk. It's not that I had kept every single piece of scrap paper (although I came close), but I definitely had a *few* large bins' worth of crap to sort. Whether you have kids or you still have your own notebooks from high school or college, it's worth going through everything and finally letting go!

The daunting aspect for me was the sheer volume of things to sort through. And the excruciating part was the decision-making fatigue, because every single thing I picked up required me to make a choice about its fate. Each item raised these questions: *Should I keep this? Should I donate it? Is it trash? Does it get recycled?* The items that really tugged at my heartstrings were notes that friends had passed to me in high school between classes, including a few handwritten notes from my best friend, who passed away when I was in Bali. Paper items like that are hard to throw away, but one great way to hold on to paper items while reducing physical clutter is to scan them and store them digitally instead.

Ultimately, I let most of my school supplies and textbooks go to the donation center. I shredded anything with personal information and recycled the rest. I realized that I wouldn't take the time to reread or reuse most of my school things, so I thought, *Why bother letting them take up precious space in my home?* And my best friend's notes? I kept those.

Where to Declutter

That's simple! Easy for me to say, right? Once you've chosen *what* you want to declutter first, the next-most-important step is to decide *where* you want to do the work. It's probably best to work in the space where your chosen category of items lives (or *should* live). With the paperwork example, the home office makes sense. If you're dealing with tools, then your shed, garage, or tool closet would probably be most appropriate. Here's a list of places that you may want to consider decluttering (add anywhere else that's relevant to you):

- Attic
- Basement
- Bathrooms
- Bedrooms

- Closets
- Dining room
- Entryway
- Garage
- Kitchen
- Laundry room/area
- Living room
- Mudroom

- Office (at home and/or at work)
- Parents' house
- Playroom
- Pool house
- Shed
- Storage unit/locker
- Vehicle(s)

When you review this list and think about the areas in your home and in your life that are applicable to you (no pool house? me neither!), think about where you'd like to start. Is there a specific part of your home that could benefit the most from decluttering? Perhaps a place that you feel the least comfortable, or the most stressed, because it's overflowing with stuff or is seemingly impossible to organize? If that sounds like your entire home, you might really have your work cut out for you.

While some people might recommend starting with the easiest room, I say begin with the area in your home that you feel needs the most attention. You can still start with the "easy stuff" in a "challenging space," because your first small accomplishments will kick-start your motivation to get through the tough stuff. Regardless of where you choose to start, the most important thing is *that you start*. Just as with anything else, action will be the most effective tool to get the job done.

Assess Your Stuff

Once you're done sorting, take a moment to reflect on what's in each pile. The best way to reveal your buying habits is to take stock before you make a move to get rid of everything. Maybe you have a sock-hoarding problem you weren't fully aware of. Or maybe you'll discover twenty brand-new tubes of toothpaste in the cupboard that you never gave any thought to before. The process doesn't have to be formal. It's really about taking mental notes to recognize the patterns in your purchases. A big part of becoming a conscious consumer is to understand your current consumption habits. So ask yourself:

- What items am I collecting?
- Do I have multiples?
- What things are going most unused but appear in abundance?
- Why am I buying so much of this one item?
- Do I need to continue to buy this item?
- Do I need these collections? Will I use them? Do I love them? Do I find them beautiful? Do they serve this season of my life?

When I did this audit of my stuff, I discovered that I had developed a habit of buying nail polish. It's not like I had hundreds of bottles, but I'd say my collection had at least twenty-five to thirty bottles in its most plentiful state. *That's not so bad,* you might be thinking. And you're right, it's not—except I only painted my nails a few times each year. When I stopped to think about it, I realized that I wasn't overly attached to my nail polish collection. How many shades of pink did I really need? Whereas, if I knew I was in love with my collection, I wouldn't have given it another thought! I'd suggest the same to you. Collections are great if you really love them, and if you do, then keep them!

I decided to give away my collection to a friend who gladly took it off my hands, and now I go get a manicure or pedicure when the mood strikes, which isn't often. The best part? I no longer feel the urge to drop fifteen to twenty dollars on a bottle of nail polish when I'm at the store. When I gave away those bottles that day, I made a decision about the person I was going to be in the future: Now I'm just not a person who buys nail polish. This is your opportunity to question your habitual purchases and to make conscious choices about what your actions will be in the future—to break the cycle and create new habits.

As you go through the items that you're sorting and decluttering, this is also an ideal time to reflect on the life cycle of your things, including how and where they were made. Consider the following questions:

- Where was this made? Is this product local or imported?
- Was this made ethically and sustainably?
- What materials were necessary to make this?
- Could this have toxic chemicals in it? (I recommend watching the documentary *Stink!*, about common household products with toxic chemicals.)
- Can I repair this or have it repaired by someone else?
- Could this easily be recycled?

These questions aren't intended to make you feel bad about what you already own and what you buy. They're meant to give you some context around your previous purchases and to help you set your values for acquiring new things moving forward. If you can get in the habit of asking yourself these questions when you buy something new, then you may either reconsider the purchase entirely or check to see if you can get a sustainable, ethical, and high-quality version of that item.

I'm by no means perfect when it comes to the questions on this list. Not everything I own or everything I've purchased in recent years is an ethical or environmental victory. But I do find that these questions have helped me buy less and buy better, while finding options to shop secondhand or borrow and rent where possible. To me, the connection between minimalism and waste is palpable. The less we consume, the less we waste. Even if my purchasing history isn't perfect, I don't foresee any more storage units in my life, and that in itself is a big step forward.

The Connection Between Minimalism and Waste

Often we purchase way more than we need. It's easy to overshop because, well, stuff is cheap, thanks to overseas production. When we buy a new product, whether it's clothing, furniture, decor, or a new cellphone or car, that product requires resources in order to be made. Those resources need to be extracted, refined, and transported. From there, products need to be manufactured, packaged, and shipped. Unfortunately, the reduced prices we pay have come at devastatingly higher costs for people, in terms of labor conditions, income, and health, and the planet, in terms of habitat loss, pollution, and waste. Each time we make a purchase, we're supporting that system.

Take a simple cotton T-shirt as an example. As Annie Leonard highlights in the book *The Story of Stuff*, we can buy a cotton T-shirt for as low as $1.99, but cotton happens to require a *lot* of water and use a *lot* of insecticides (contaminating the surrounding natural environment), which also impacts worker health among cotton growers (causing neurological and vision disorders).[1] As Leonard further explains, turning raw cotton into fabric is also energy intensive, and even the dyeing process is chemically demanding; when the cotton is ready, it's shipped off to be sewn into T-shirts, which often happens in a sweatshop to reduce the costs to the brands buying them.

In total, that cotton T-shirt requires about five tons of CO_2 to produce, and that's not including transportation. I know I can't look at a cotton T-shirt the same way again. Can you?

So take a moment. Let it soak in. And then go count all the cotton pieces you own. Once you've done that, try taking an even broader look at everything in your home. Confront your stuff—not out of guilt but out of a desire to learn and to change. Each and every single item you have acquired over time has required vast inputs of resources, production, and transportation in order to land in your drawers, on your shelves, or in your cupboards. This snapshot doesn't even touch on the issue of textile waste—something I'll address in much greater detail in chapter 6, "Outfit Repeater." However, the T-shirt example does show that the process of making stuff creates a lot of waste and pollution, in addition to having other negative impacts on the natural environment and human health.

As I began to understand how a minimalist mindset was essential to a sustainable lifestyle, I became convinced that I could make a difference in this world by becoming a more conscious consumer (see chapter 3). In other words, it was important to me to avoid buying cotton T-shirts for $1.99, because I didn't want to support a system that was heavily polluting the planet and unhealthy for people.

After reading this section, you might be feeling a bit less inclined to let go of your things, considering how much energy and resources went into making the stuff you already own. Don't let that stop you from clearing out your unwanted stuff. You know that saying: One person's trash is another person's treasure. If your possessions can be rehomed, do it! Don't let guilt get in the way of your simplifying your life. This is your opportunity to start fresh on your journey to becoming less trashy, equipped with a new and minimalist lens through which you can view the world.

Responsibly Unload Your Crap

This wouldn't be a book about living less trashy if I told you to declutter your stuff and then *throw it all out*. That would be awful advice. My suggestion? Don't be trashy. You made your sorting piles (keep, sell, donate, recycle, and trash), with trash being the last resort. Inevitably, some of your stuff will be trash. In that situation, don't feel guilty about throwing something in the garbage; either it's sitting in your house as trash or it'll be trash in the landfill. So just throw it out and let your guilt go with it.

How you manage the rest (apart from your "keep" pile) is just as important. There are so many great options to rehome your unwanted things.

Selling

In my experience, it's best to try to sell only high-value items, as I've found it mostly tedious to attempt reselling low-value things. What's worth your time to sell and what's not is up to you and might depend on external factors, like how much you want/need the extra cash. Decide what your time is worth! Maybe you want to assess it in terms of an hourly wage: If it takes an hour to sell something, but you won't make your hourly wage on the sale, then perhaps it's not worth your time. For me, it was definitely worth my time to sell my DSLR camera. Even though it was a few years old, with a quick Google search I discovered that it was still worth several hundred dollars, and I sold it quickly online. Unwanted home decor items, however, didn't sell very quickly and I didn't make much cash back. When it comes to clothes, I take any of my higher-end, good-quality pieces to a consignment shop, and the rest gets donated. If you have some stuff that you've decided you want to sell, you could try some of these options:

- **Garage sale:** It may be old school, but it's an option to consider. You may need to put in a bit of time and effort if you go this route, and garage sales don't always yield big payoffs, but if you gather some friends and family, you might find you have a lot of fun. Garage sales are a good venue to sell just about anything you own.
- **Consignment:** Consignment shops accept used items and sell them on your behalf—and you'll earn a small percentage of the sale once the item is sold. There are consignment shops for everything from fashion to furniture. You can consign locally at brick-and-mortar shops and find some consignment options online, too, like thredUP, Poshmark, and the RealReal!
- **Online marketplaces:** Craigslist, Kijiji, and local Facebook groups are great venues for selling your stuff. Keep in mind that they typically require in-person transactions. You can also sell on websites like eBay (this type of selling will likely require you to package and ship items).
- **Pawnshops:** These are typically immediate transactions where the pawnshop offers you cash on the spot if they are interested in buying what you have to offer, from video games and jewelry to electronics and musical instruments.

This is by no means a comprehensive list for selling your stuff, so don't be afraid to get creative. Where or how you sell your stuff really depends on what you're trying to sell. If you have a bunch of scrap metal lying around on your property, you might be able to get some cash for it at a scrapyard. There are also a lot of online marketplaces specifically dedicated to used electronics (like ItsWorthMore.com) and antique markets for vintage goods (like Etsy or eBay). The best way to know if you can sell your unwanted items is to do a few online searches to find out if there's a market for your things.

Donating

Through decluttering, you may have discovered some things that are still in good condition but that you no longer love and aren't worth your time to sell. If you find yourself in that situation, here are some ideas of where you can donate those items instead:

Tools: Find out if you can donate them to a local tool library.

Construction materials: Consider donating to Habitat for Humanity's ReStore or a similar organization.

Clothing: Find out if there are any shelters, food banks, or other nonprofit organizations that will take clothing, shoes, and accessories in good condition.

Books and magazines: Find out if your local library will accept your donations. Your community may also have Little Free Libraries (little freelibrary.org), with the opportunity to donate a book or magazine there. Used-book stores will often accept titles in exchange for cash or a store credit, and there may be a charitable organization doing a book drive near you.

Art supplies and crafts: Call up local schools or childcare centers to find out if they will accept these types of donations where you live.

Toiletries and miscellany: From shelters to churches, mosques, and related nonprofit organizations, there are a number of groups and institutions that help people in need. Find out if anything you have can be donated there. These products should be unopened, unused, and not expired.

Food: Food banks, shelters, and other nonprofit organizations may take your unwanted food items. (Ensure that your items meet their requirements for donations.)

Just about anything: While not technically a donation, a great way to get rid of your things is to offer them for free to your friends, family, and neighbors. I personally find this to be a very successful approach; I take a quick photo with my phone and send it to people I think might be interested in the item(s), and they let me know if they'll take it or not. You can also give stuff away for free online with marketplaces like Craigslist and Kijiji and Buy Nothing groups on Facebook.

It may seem like a big task at first, but if you stick with it, you'll likely be able to find a new home for most of your unwanted things. Alternatively, you could make a call to Goodwill (or a similar charitable organization) and have them come by to pick up *all* of your decluttered stuff. If you're already juggling a demanding schedule, this option makes it as quick and painless as possible. But if you do have the time to spare to research where to sell and donate your things, you will help ensure that each item finds the best home possible. Why does that matter? When you match your stuff with a good home, you reduce the likelihood that it will end up in the landfill. Goodwill puts a lot of effort into selling our unwanted stuff, implementing systems to make the landfill a last resort. Unsold clothes, for example, are often sent to textile-recycling organizations, but if clothes are wet, moldy, or contaminated, they go directly to the landfill.[2]

Decluttering will likely be ongoing as your life changes, but the lessons you learn through this big clean-out will help you in the future. If you simplify your life by buying less and continue to ask yourself some of the questions in this chapter as you declutter along the way, there won't be another massive decluttering endeavor in your near future, if at all. Plus, with this book under your belt, you'll likely be seeing the world through a new lens, one that reveals how everything we consume is connected to waste. On that note, let's jump into ways you can avoid recluttering, so you can maintain the fruits of your labor in the long run!

Develop Tactics to Avoid Recluttering

The point of this chapter is to help you start fresh, streamline your home, get rid of your unused and unwanted stuff, and work toward a simpler lifestyle. At the same time, the task of decluttering gives you the opportunity to examine and modify your consumption habits. You don't want to end up adding more stuff to your space after you did all that work, right? In order to make sure your home stays as neat and tidy as possible, now is the time to develop the minimalist habits that will help you

- avoid recluttering;
- save money (because you'll buy less); and
- reduce your waste (because you won't cycle through products so quickly).

Remember those bottles of nail polish I used to collect? If I had started buying more after giving my collection away, I would have defeated the purpose of the whole decluttering exercise. Instead, I made a commitment to myself not to buy new nail polish bottles and kept it.

It's easy to make a resolution—most of us do it at least once a year, when the calendar flips—but it's much harder to follow through on the commitments we make to ourselves. Before you make a commitment, ask yourself the following questions, and make sure you come up with some really solid answers.

- What do I want to get out of all this work?
- What do I have to do to get there and maintain that goal?

Remember your Minimalist Mission Statement? Well, these two questions will help you come full circle to where you started on this journey to living with less. After some introspection you will learn more about yourself and set yourself up for success in the long run, providing yourself with an action plan for possible moments of relapse.

Through your decluttering, you'll likely have discovered your weak spot for collecting books or buying jackets, or maybe you feel the need to purchase every kitchen gadget ever made. Know what your weaknesses are and set boundaries and rules for yourself to avoid falling prey to your desires (or to ads). While you're at it, take time and make room for gratitude. Be grateful for what you already own, and don't let our consumer-driven society fuel your need to buy more. Pause and reflect on what you have already, on what you love and enjoy about your life, and let those positive feelings help you stay the course. While you're at it, write down or take mental note of the things you love the most that aren't actually *things*. Keep this list in mind when you're feeling the urge to shop, and instead replace that urge with the non-things you love. The following are some tips to help you avoid impulse shopping.

Curb Your Impulse Shopping

Are you an impulse shopper? Do you buy things on a whim that you didn't plan or budget for? Making unnecessary purchases can be a financial drain and can create a lot of waste. Those Amazon boxes and bubble wrap add up! Here are a few tips to help you avoid buying things you don't need. Make sure to buy what you need, when you need it, without falling prey to an advertisement's promise of a better life.

Keep a wish list: This is my all-time favorite tool for managing my consumption. On my phone I keep a wish list of items I'd like to buy. If I find something that interests me that's not on my list, I'll add it and give it some time before I make the purchase. Having some time to "sleep on it" will give you the chance to think it through and decide if you really need it. Chances are you'll end up removing it from your list, and you'll be thankful you didn't buy it on impulse. On the other hand, if it sticks around and you know your life could really benefit in one way or another from making the purchase, then you know it's likely a worthwhile investment. If it helps, add a time frame (like ten days, two weeks, or a month) during which you don't allow yourself to make the purchase, and after that period of time, decide if you'll buy the item or delete it from your wish list. Some items stay on my wish list for months, because the purchase doesn't make sense yet. For example, a new pair of Birkenstock sandals are on my wish list, but I won't buy them during winter because I won't wear them until late spring.

Avoid your known triggers: If walking through a shopping mall is likely to inspire random purchases, avoid the mall. If you must go for something on your wish list, be sure to get in, get what you need, and get out ASAP. Similarly, don't spend hours online scrolling through your favorite products. Window-shopping online can be just as powerful a pull to spend money as shopping in person.

Reduce exposure to ads: Whether we're on social media, watching TV, or flicking through magazines, we are bombarded by ads all the time. Limit your exposure to ads where you see them most, even if it means unfollowing your favorite fashion brand on Instagram!

Buy for your real life: Remember those high-heeled boots I bought that were perfect for life in New York City? I didn't buy those for my real life, unfortunately. Figure out what you tend to buy for your fantasy life, and face the reality that perhaps those things aren't for you. For example, if you spend time and money collecting art supplies, but you don't actually spend time creating art, maybe art isn't your true hobby or passion. Similarly, if that guitar is collecting dust, don't go shopping for another one, even if it's on sale!

Pursue hobbies you enjoy: Remember to also live your real life and pursue your actual hobbies, especially the ones that don't involve shopping. Minimalism isn't about living an austere life. Go hiking if that's your jam! Get together with friends and family for a backyard barbecue. Explore a cute town nearby. Become an expert knitter. Volunteer with a local community organization. Do the things you love, and let your real priorities fill the gap that shopping used to take up!

Develop your sense of style: This tip is just as important for your wardrobe as it is for the rest of your home. Instead of following the latest must-have trends, develop your own timeless sense of taste and style. Your closet and your home should be about *you* and what *you like,* and everything you own should tell *your story*. When you create your own style, you'll naturally shop less because you won't be chasing trends. Spend some time flipping through magazines at the library or search online for ideas to develop your personal style guide for your fashion and your home decor.

Set up rules for yourself: If you tend to like more structure, throw in a few rules that you know you can follow. Maybe you'll decide to live by a "one item in, one item out" philosophy (which doesn't necessarily reduce consumption but helps keep things minimal). Another idea is to follow a budget; make sure you have enough money saved up to make a purchase, and don't allow yourself to buy it on credit—we'll chat more about finances in chapter 10!

Ask yourself some questions before you make any purchase:

- Do I really need this?
- Do I have something at home already that serves the same purpose?
- Will I want to maintain and repair this?
- Could I borrow or rent this instead of buying it?
- What else could I use this money for?
- Do I have the money to spend on this right now?

I love this quote from The Minimalists: "Everything is 100% off if you don't buy it."[3] If you're a sucker for sales, that's the line you need to remember! We often hear or say things like "Oh, I saved so much money today because they had such a huge sale!" But wait, did you *save* money, or did you *spend* money? If you spent money, then you didn't really save anything.

Reflection and Checklist

I didn't become a minimalist right away, and I wouldn't even say that I'm a minimalist now. I have, however, chipped away at my surplus of stuff over the years. At the same time, I was exploring what it means to live more sustainably, and I couldn't help but see very significant parallels between the two movements. In a nutshell, minimalism helps facilitate a sustainable lifestyle: The less stuff we consume, the fewer resources we use up.

Minimalism doesn't happen overnight, and it's up to you to decide what it means to you and what it looks like in your home and in your life. Decluttering your home can take weeks or months, depending on how much of a priority you choose to make this activity. Want to be a minimalist badass? Do it! Go clear that clutter and refuse all the new things, while eagerly analyzing each and every object vying for a space in your home. Even if you're not gung ho on clear countertops and empty drawers, everyone could use a little tidying up in their lives. Define what that means to you, set your goals, create your new routines, curb your consumption, break up with the Joneses, and decide on what your new normal looks like. Your stuff tells a story about you. Make sure you write that story, not the marketers, advertisers, or the Joneses.

While this chapter is focused a lot on *stuff,* the act of decluttering stuff and moving toward a minimalist lifestyle is actually meant to provide the chance to focus on everything *but* stuff. By giving ourselves more breathing room in our homes, and by spending less time consuming more, we can direct our attention to the things in life that bring us joy. This is your chance to figure out what those things are! What's most important to you? Is it family time? Volunteering? Visiting with friends? Pursuing your hobbies? Starting that new business? Writing a book? Innovating? Figure out what it is you want to do with your extra time, and perhaps even your extra cash! Your minimalist journey helps lay the foundation to craft your best life.

This chapter may seem like a lot of work, but it's best not to look at it that way. It's about taking it one step at a time. For simplicity, here's a quick recap of the activities listed in this chapter:

- Create your Minimalist Mission Statement
- Take note of things in your home that you love
- Take note of things in your home that have gone untouched and un-loved for years
- Think about everything you own and ask yourself: *Why do I have this?*
- Declutter your unloved and unused possessions
- Find the patterns in your own consumption behavior
- Discover the connection between consumption and waste by assessing your belongings based on their environmental and human impacts
- Find lovely new homes for your unwanted things
- Get to know your shopping habits and learn what you tend to buy most often, then shop less
- Set boundaries for yourself to avoid recluttering, and curb your consumption

Chapter 3

Conscious Consumption

Buy less and buy better

No Logo

Pink hair and cargo pants. It's nostalgic and occasionally embarrassing to reflect on my fashion choices when I started grade nine. I went to high school in the early 2000s, a time when Blink-182 and Destiny's Child were the CDs in my Discman, and way-too-baggy pants were cool. My go-to outfit was a pair of baggy low-rise pants, probably khakis, coupled with a spaghetti-strap tank top. I finished off the look with skater shoes and pink highlights in my hair, channeling Avril Lavigne for that punk-rock vibe. Did I skateboard or play the guitar? Nope! I was a total wannabe, obsessed with brands like Roxy, Quicksilver, and Billabong. In other words, I was totally branded. The brands I wore told a story about who I was, or at least who I aspired to be.

Throughout high school I wanted to be a fashion designer. It wasn't a passing desire, like tossing around the idea of becoming an astronaut while being crap at math. I actually worked toward this path, taking courses that taught me to design and sew garments, and I got pretty good at it. Watching fashion shows on TV was totally my jam, and I aspired to be one of those well-known fashion designers with their fresh looks strutting down the runways of New York, Paris, or London. I soon learned, though, that the fashion industry wasn't as glamorous behind the scenes.

When I entered high school, buying clothes wasn't about checking the label to find out the country where my T-shirt was made. That nonchalant attitude changed as Naomi Klein's debut book, *No Logo: Taking Aim at the Brand Bullies,* became an international bestseller and everyone at my high school started reading it. Published in 1999, this book was a game changer for the antiglobalization movement, as it took us behind the scenes to the operations of large, multinational brands. Klein uncovered all their dirty secrets: youth-centered marketing and the purposely limited product choices available to consumers; outsourcing of manufacturing to less developed countries, in which working conditions have been documented as abysmal, with payment below the poverty line; destruction of the local ecosystems, impacting both communities and the natural environment. Sadly, these conditions still persist as I write this book.

Toward the end of my graduating year, I was accepted into college for fashion design in Toronto. I decided to turn it down because the industry, from what I understood, was too toxic. *No Logo* had filled my head with visions of sweatshops and child labor, and rightly so. While I can't say that I finished high school with a timeless sense of style (I was still rocking those skater vibes), my newfound outlook on the fashion industry gave me a conscious perspective as a consumer.

It took a while before any reasonably stylish clothes were made sustainably and ethically. The slow and sustainable fashion movement is still in its infancy, even now, but at least there's a lot more available on the market. After reading *No Logo,* I remember searching the internet to find fashion brands that aligned with my values, because there definitely wasn't anything available at my local shopping mall. Unfortunately, it was slim pickings online, too (the internet and e-commerce were also in their infancy). A lot of the more ethical/sustainable options available at that time didn't reflect my punk-rock taste, falling more on the hippie end of the style spectrum instead. Thankfully there's enough variety available now to indulge your style, whatever that may be.

I'm grateful that more and more brands are socially and environmentally conscious. I believe it's important to vote with our dollars for a system of production and consumption that supports ethical (fair treatment of people) and sustainable (fair treatment of the planet and its resources) business practices. While I believe we need to do more than be conscious consumers to make a positive impact in the world (we'll chat more about this later), we should still hold all brands, in particular multinational companies with deep pockets, accountable for their actions. Supporting conscious brands and avoiding the rest whenever possible can help steer us in the right direction.

Being a conscious consumer isn't just about buying sustainable and ethical fashion. It's about being mindful of the life cycle of *everything* we consume, from fashion and furniture to electronics. This chapter is a crash course in conscious consumerism. Here's what we'll cover: why our current system of consumption sucks, and what we can do instead; how to buy less and buy better; defining what *better* means to you; and engaging in accountability.

Like everything else in this book, this lifestyle change won't happen immediately, and it's not about being perfect, either. The point of this chapter is to help you shift your focus toward supporting a better system. Whether we realize it or not, we're participating in the system, and this journey helps us decide how we believe the system should operate—and then act according to those beliefs. Personally, I want to support a system that uplifts the people involved in the production process of the goods we consume and that doesn't leave a wasteland in lieu of nature. So I seek out businesses that align with those beliefs and support them with my dollars. This chapter outlines how you can do that too, based on your beliefs and preferences.

Why Our Current System of Consumption Sucks, and What We Can Do Instead

Our economy is built on consumption. Not just a little bit of consumption; it's more like our planet is one big factory operating to feed the profits of a handful of big corporations, and we all need to keep shopping in order for it to function. Brands collectively spend billions of dollars on marketing to make sure that we buy at unprecedented levels. Then, when we hit an economic downturn, we consume less, which decreases the number of jobs available. The economic system falls apart when we stop consuming, because so many people end up out of work, unable to pay their bills or put food on the table. You can see this pattern in the Great Depression (1929–33), the Great Recession (2007–9), and the COVID-19 pandemic (started in 2020), among other notable economic downturns. It's devastating that so many people are unable to attain a basic quality of life during these downturns.

The current economic system is set up so that we, as consumers, feel insufficient. When we feel like we don't have enough, we get an urge to acquire more to fill the void. We're like hamsters on a wheel, consuming and consuming and consuming. It's a never-ending cycle called the hedonic treadmill; no matter what we get or achieve, we still want more. We think to ourselves, *If I can just buy this one thing, then I'll be happy.* Can you relate? I know I can. Maybe it's a new cellphone or a designer handbag. The problem is, once we purchase something we want, we aren't left feeling fulfilled. There's often something new that we set our sights on. We got the new phone, but now we want a high-end watch. When does it end? It doesn't, and that's the point. The scheme is set up to make us continually go after the next thing.

Let me be frank here; I'm not against consumption. One could argue that to be human is to consume. And I agree. It's our very nature to alter our surroundings to consume food, to create a habitable space, and to enjoy life. I'm no different. I eat food. I like a nice and cozy home to live in. Preferably a home that heats in winter, cools in summer, doesn't leak, is mostly spider free, and has running water and comfortable furniture in it that I find beautiful. These are luxuries, of course, because not everyone has access to such amenities; it's definitely a privilege. I'm a consumer, and we're all consumers to varying degrees. I'm just saying we can do better; we can be better consumers to support a better system. And we shouldn't have an economy that depends on our consumption alone to ensure that people have jobs, homes, and food.

In a world with nearly eight billion people, this consumer-driven lifestyle is not sustainable based on our finite resources. There's nothing wrong with financial wealth (that's covered in chapter 10), but how we live our lives and how we choose to spend our money does matter. Ask yourself this: *How much of what I have and what I want (in terms of both material things and lifestyle) is based on my own true desires?* Think about how you respond to this question. It can be quite revealing and even complements the exercises you did in the previous chapter on minimalism and decluttering. You might discover that you're chasing someone else's dream and not your own.

Earlier we discussed the planned obsolescence of style, in which products are designed to be trendy and quickly become outdated so we consume more. Many of us have become quite aware of this cycle in the fashion industry, with new trends coming out on a weekly basis. But it's not just fast fashion that has us trying to keep up with the latest trends. Many other industries are adopting this model, including furniture and home decor. No longer are we living in an era of high-quality, locally produced couches and bedroom sets that we hold on to for generations. Most furniture manufacturing has been shipped overseas along with production of the rest of our products, resulting in lower quality, cheaper prices, and faster-changing trends.

We kick our old furniture to the curb quickly and easily because it wasn't a big investment in the first place. It's often not worth packing for a move, and I see that firsthand because I live in a university town. At the end of each school year, students leave their cheap furniture at the curb as they move on with their lives. It has created a big trash problem for my city to deal with, and it shows our lack of interest in caring for our belongings. It's sad to see a decent couch sitting in the rain because no one cared enough to find it a new home. Then again, people typically don't want secondhand fast furniture, because they can go buy it brand-new so cheaply.

It's hard to win against a system that benefits from planned obsolescence and cheap stuff. When these cheap products get made abroad, we often have no control over the extraction of resources (is it sustainable?), the treatment of people (is it ethical?), or the quality of materials being used (are they durable?). Additionally, the shipping requires energy, transportation, and packaging. This, in my view, is a broken supply chain and a system that doesn't support people or the planet.

A quick note about price. There is a valid criticism that sustainable and ethical products are expensive, elitist, and unattainable for the poor. While I do agree that this is a challenge to overcome, let me play devil's advocate. Recall that cheap stuff has literally been produced by the sweat and labor of the poor overseas. And that all products we can find so cheaply today used to be expensive back when they were locally produced. People used to save for their purchases and frequently shopped secondhand! These days, cheap new stuff is exploitative of the global poor and traps us in an endless consumption cycle with credit card debt, and yet we see high-quality and durable items as symbols of privilege? I believe that these criticisms can, unfortunately, distract us from creating solutions that uplift *all levels of society.*

The life cycle of our material possessions has one guaranteed final destination: the dump. But thankfully, there's an alternative. We can support a circular economy, one in which extracted and raw materials are actually valued, and they're used to build products that are high quality, have longevity, and are repairable. At the end of a product's life, the raw materials can be once again extracted and reused to make new products. For example, did you know that electronics contain valuable precious metals like gold, silver, and platinum? And that the process of extraction for those raw materials is very dirty, energy-intensive, and polluting? Supporting a circular economy promotes the reuse and recycling of materials, so that they remain in the system instead of heading for the dump.

Here are some tips to help you support a circular economy and a better consumer system:

- Demand repairability of products (either by communicating directly with brands or by supporting government policies for repairability).
- Buy products that are easily recycled either by design (the fewer types of materials fused together, the better) or because the company you're buying from will take back the product for use in future products (a total win!).
- Become an engaged citizen. Communicate with your government (at all levels) and the companies you buy from to support sustainable and equitable systems. You can do this through engagement, advocacy, policy support, and your purchasing power. Where do you want your personal and tax dollars to go? Make sure to have your say by showing

up at public meetings, writing to your politicians, voting, and adding your comments to new policies, legislation, and government-led initiatives.

- Find the topic you're most passionate about and pursue it specifically and directly. Whether it's humane animal treatment, organic agriculture, or the right-to-repair movement, focus on a specific element of conscious consumption that you're keen on and make a difference there. For example, I have experience with environmental land use planning from when I was a professional urban planner. Saddened by the loss of natural areas (like forests and wetlands) due to land development (another form of consumption), I support nature conservation directly through donations to charitable organizations. As individuals, we can't directly and actively support all issues all of the time, but we can join others to work toward improving specific causes that we are passionate about.
- Keep in mind that the price tag doesn't represent the full and true cost of an item you're purchasing. The true cost may include biodiversity loss, pollution, low wages, and or harsh labor conditions. This way of thinking isn't intended to pile on the guilt; it's intended to be a reminder to look at the bigger picture and to motivate you to create systemic change for the better.
- Support the secondhand economy. Thrifting is such an important means for us to keep useful products out of landfills. Plus, it's way more affordable than buying brand-new!

How to Buy Less and Buy Better

The best way to curb consumption is to not buy anything *at all.* Did your jaw just drop? For most people, quitting shopping cold turkey might not be realistic. I've done it myself, and I share more about that experience in chapter 11, "No-Buy Month"; I'll encourage you to try it out for yourself for a month—or maybe even longer. But don't worry, I'm not saying that you have to live without toilet paper if you run out; there will be no dirty bums because of this challenge! It was a purposeful challenge to not buy *unnecessary* items like clothing, shoes, home decor, and any other miscellany.

Buying less is not about self-deprivation. When you buy less, it saves time (you spend less time shopping), it gives you the opportunity to focus on more important things (like friends, family, hobbies, etc.), you save money or have the chance to pay down debt, and you're less trashy (buying new things uses resources, creates pollution, and results in waste). The benefits abound, but chances are, you'll still end up shopping at some point. Let's chat about how we can choose to buy better when we do end up trading our dollars for stuff.

I've noticed over the years that the high-quality products I've invested in last longer and have a better resale value, even if I bought them used. These products also tend to be (mostly) timeless. This has been very true for my wardrobe. The cheap clothes I've purchased have rarely been accepted at my local consignment shop for resale, and my friends don't typically want them either. In contrast, when I've invested in high-quality clothes made from better materials with evidently better construction, those pieces remain in high demand even if I'm no longer interested in wearing them. Remember those knee-high cognac boots I discussed in the previous chapter? They were fantastic quality, made in Italy, constructed with gorgeous stitching, and had a flattering fit. Even two years later (unworn, sadly), my local consignment shop had no problem reselling them. They were stunning, but they just weren't for me. Even my worn items, when they are high quality to begin with, still have a high value when it comes to reselling. And it's not just clothing, either! This goes for all of the things we buy and own. Clearly, durable and timeless products can have a longer life and be enjoyed by more people as they get passed along, which supports sustainability and a thriving secondhand economy.

If you're just scraping by because you're broke, then you might be thinking, *Sheesh, Tara, you're awfully privileged to be telling all your readers to go buy good-quality stuff!* I hear you. And as with everything else in life, we move forward through our intentions. As you aspire to change things in your life, I believe it is worth keeping this buy-less-but-better objective in mind as you improve your personal circumstances. So if right now you're living paycheck to paycheck and this section doesn't seem terribly relevant, keep it in mind for when you achieve greater prosperity. Plus, this "less but better" framework isn't just applicable to buying new things; this approach is also relevant for thrifting and supporting the secondhand market.

When my husband and I moved into our first home shortly after getting married (literally a week later—it all happened so fast!), we didn't have much spare cash for furniture. Most of what we did have came with us from our childhoods or we inherited from our grandparents. We were incredibly grateful for the family heirlooms, as well as a set of three industrial-chic coffee tables we received from Hubby's parents as a wedding gift.

Our dining room table was gifted to me by my grandmother when she downsized, a few years before she passed away at the ripe old age of ninety-two. What I love about it, aside from the fact that it's beautiful and made from solid wood, is that it is over a hundred years old, and it wasn't even brand-new to her! It still came to me in perfect condition. It has a long history and still has many more years of family gatherings to host. What could be more sustainable than that? While we were lucky to receive these beautiful family heirlooms, we still needed quite a few pieces of furniture for our new home. So I went on a mission to find them sustainably *and* on a tight budget.

Before we bought our house, we were living in an apartment with the basics. Our bed was simply a metal frame that held our box spring and mattress. There was no headboard, and nothing fancy about it. While it wasn't going to be featured in *House & Home* magazine anytime soon, it got the job done. Intent on making our new house feel like a home, I searched eagerly for a beautiful bedroom set that could bring joy and comfort into our space. A quick Google search made it quite clear that a full bedroom set made from real wood would cost *thousands* of dollars (at least the stuff I was looking at; my husband always tells me that I have expensive taste). Undeterred, I moved my search to Kijiji. Immediately I found what I was looking for! Within a forty-five-minute drive we could pick up a full bedroom set, including a queen-size bed (with a headboard!), two side tables, a dresser and mirror, and an armoire for a mere three hundred dollars. Sold.

The family we bought this set from was upgrading their bedroom because they were looking for a refresh. As the saying goes, their trash was my treasure; the entire set was made from solid wood, and it's still our bedroom set to this day. We get so many compliments on it, and I'm always proud to share the story of my good-quality find at a price we could afford to pay. This set is probably about sixteen years old now and still going strong.

The more we can keep out of landfills, the better. How do we do that? With the Less but Better Method.

The Less but Better Method for Shopping

Hopefully by now you're convinced that becoming a conscious consumer is a worthwhile endeavor. New habits tend to be easier when there's a framework for us to follow. The Less but Better Method is the approach that I use for shopping, and it's one that you can easily adopt, too! Here's how it works: Whenever you have a purchase to make, you are confronted with a choice. Either you can choose to buy something that's less expensive and cheaply made, which likely means you'll wind up replacing it multiple times over the course of your life (and probably spend more money and send more to the landfill), or you can purchase something that's high quality to begin with, that you can repair and maintain and hold on to for much longer (resulting in less waste and more savings in the long run). Whenever possible, I opt for the latter alternative. But there are a few other important considerations before you decide to make a purchase—like asking yourself if you even need to buy something in the first place. To apply the Less but Better Method to your own life, consider the following options before you purchase.

Keep a wish list: Back in chapter 2 you were introduced to the idea of keeping a wish list of items you want to acquire, with the purpose of curbing your consumption. Your wish list will (hopefully) help you focus on the *less* part of the Less but Better Method.

Use what you have: When you consider what's on your wish list, think about how much you actually need each item you have listed. Take a moment to consider if you could use what you already have. Like, can your chef's knife do the job of cutting that pizza, instead of your going out and buying a specialty pizza cutter that you know you'll rarely use? Put everything through this test.

Borrow it: Friends and family might be able to lend us exactly what we need when we need it. Similarly, libraries, tool libraries, and other free resources may exist in your community that will let you borrow things like books, tools, kitchen supplies, etc. for free or a nominal membership fee. Consider if you can borrow instead of buy.

Rent it: If you can't find what you're looking for among your friends and family, and it's not available to borrow from local resources like a tool library, find out if you can rent it. Take a carpet cleaner, for example. It's a specialty item that we don't need to use often and tends to be expensive to buy. Find out if it can be rented at a local hardware store, or even online! Think bigger here, too. Like the time my husband and I rented a fully equipped trailer for a localish road trip using a website called Outdoorsy that connects recreational vehicle owners with would-be renters. Consider that you might be able to rent just about anything you want; you just have to do a quick search to find out if it's an option. This is called the sharing economy, which facilitates peer-to-peer transactions and reduces the need for ownership. Cool, right? You may be familiar with more well-known models for this, like Uber and Airbnb. Why own when you can rent? I'm certainly grateful I don't have a trailer sitting in my driveway taking up space and requiring maintenance! (On the flip side of this, you can rent or loan out your specialty items and earn some extra cash!)

Swap it: Think of swapping like a garage sale, only everything is free, and your closest family and friends have brought all their best (but unwanted) things! Consider hosting or attending a stuff swap, and focus on one theme, like fashion, kitchen items, baby clothes, kids' toys, plant clippings, home decor, etc. Swapping doesn't have to be an event; it can also happen online through Facebook groups or similar platforms like Bunz, an online community where people can trade things like clothing, furniture, and art in real life.

Find it for free: There is so much free stuff out there! It's amazing what people are offering and what people are willing to take. A close friend of mine has had many positive experiences with swapping and free stuff: "It's given new life to lots of things I was going to throw in the trash. I posted much of it for free thinking no one would want it, and everything disappears." Similar to swapping, free stuff can come from family and friends, online Facebook groups like Buy Nothing groups, as well as online marketplaces like Craigslist and Kijiji or networks like Freecycle. You'll even find free stuff in unexpected places, like on people's curbsides.

Make it yourself: Are you handy? Crafty? Creative? If you're a maker, consider making what you have on your wish list, from art to fashion to furniture to home decor. Whether you're crafting it from scratch or upcycling—making existing products, materials, or perceived waste materials into new products—there are so many options to make things. It's a great way to divert waste and save money while getting creative and having fun!

Thrift it: Once you've exhausted the above options, shopping secondhand is an affordable and sustainable approach to satisfying your wish list. Just like I mentioned above, when I found my dream bedroom set made from solid wood for three hundred dollars, I saved money (thousands of dollars) and it was a more sustainable option (the set was already made). Secondhand goods are more sustainable simply because the necessary resources have already been extracted, the production process completed, and shipping satisfied (unless, of course, you're thrifting online, in which case shipping is still a component). The best part is, buying secondhand means you're giving an old item a new life and keeping it out of the landfill. Head to your local Goodwill, to consignment shops, or online to Kijiji and Craigslist or apps like VarageSale and secondhand e-commerce shops like eBay and thredUP.

Buy it better: If you can't find what you want for free or secondhand, then the last option in the Less but Better Method is to buy it better. What does that mean? It means to make purchases, when possible, that are better quality in terms of durable materials, excellent construction, easily repairable; made ethically, with fair working conditions and wages through the entire supply chain; and made sustainably, with the planet in mind from start to finish. Don't want your new stuff to end up at the dump? Here are some key questions to keep in mind when making your next less-but-better purchase:

- Is this item built to last?
- Does this item appear to be constructed well? Is it durable?
- Is this item constructed of high-quality materials?
- Can I easily repair this item or have it repaired?
- Does the company stand behind this product with a long-term or even lifetime guarantee?
- How practical will this item be a few years from now?
- How likely am I to still want this a few years from now?
- Are there mostly good reviews for this product?
- What are the bad reviews, if any?
- What do these reviews reveal about the product?

The Less but Better Method, if followed regularly, should help you become a more conscious consumer. But what does *better* really mean? The words *sustainable* and *ethical* often get used in marketing to help us as consumers buy eco-friendly products. Unfortunately, when these terms are used without much evidence, it can be a form of greenwashing. In other words, clever marketing is trying to sell us "green" products that aren't really sustainable once you take a look behind the curtain. To avoid greenwashing, your best bet is to do some research into the products you're buying and to make sure brands are transparent about their practices, ingredients (if applicable), and materials. If you don't have time for this kind of legwork, check out resources that do the work for you, like the website Buy Me Once, an online shop for long-lasting, sustainable products.

What I've come to realize over the years is that sustainability is truly a subjective buzzword. It can be defined in so many different ways! The best advice I can give is to define what *better* means to you, and what it means to you for brands to be sustainable and ethical. Once you develop your own framework for what it means for brands and products to be better, then you can filter out the products (and services) that fit your preferences. While it might be nice for me to do that for you, we'll likely have different perspectives on the matter. Let me give you an example.

Did you know that many synthetic materials in our clothes, like polyester, are actually made from plastic? And that the name *polyester* is just a shortened name for the material known as polyethylene terephthalate (PET), which is plastic? It has become more well known in the past few years that synthetic clothing, when washed, releases microplastics into our waterways. Like, a lot of microplastics; scientists are estimating that since the mass production of synthetic clothing began in the 1950s, about five megatons (a megaton being one million metric tons) of synthetic microfibers were emitted from washing apparel.[1] Yikes! The worst part of that is that these microplastics are making their way up the food chain[2] and have even been found in our drinking water and air.[3]

Keeping that all in mind, it has become popular to use recycled plastic as material for new clothing in an effort to work toward a circular economy. This is typically presented as a sustainability-based initiative. That said, some might argue that this approach is unsustainable because creating more plastic-based fabrics and clothing is contributing to microplastic pollution. The alternative to synthetic clothing, of course, is the use of all-natural materials like cotton, wool, hemp, etc. All of those have their own environmental footprints, too—cotton, for example, is energy and water intensive and a lot of pesticides are often used during cultivation of the crop. Once you start going down the rabbit hole here, it might leave you scratching your head and asking, *Is anything really, truly sustainable?*

Sustainability isn't as straightforward as we might like it to be. Even environmentalists have trouble agreeing what it means to be sustainable, eco-friendly, and less trashy. Debates on what is actually sustainable could go on forever! Indeed, they do, as we see these conversations play out in the news about issues like climate change. So create a list of values and elements of what it means to you for brands, products, and services to be *better,* and use the Less but Better Method as your guide.

Don't have time for that? Many people don't! Luckily, there are quicker ways to find out about whether brands are sustainable or not. While I still believe you should make a mental note of your values, or even write them down, here are a few more straightforward ways to find out if the brands you love meet your preferred standards (and to avoid greenwashing):

Certified B Corporations: "Certified B Corporations are businesses that meet the highest standards of verified social and environmental performance, public transparency, and legal accountability to balance profit and purpose. B Corps are accelerating a global culture shift to redefine success in business and build a more inclusive and sustainable economy."[4]

Good On You app: Good On You rates fashion brands based on their environmental and social performance and how they impact animals. Check out its brand directory to find out if your favorite fashion brand has been rated.

Environmental Working Group (EWG): EWG's mission is "to empower people to live healthier lives in a healthier environment. With breakthrough research and education, we drive consumer choice and civic action."[5] The key issues it covers include consumer products, cosmetics, energy, farming, food, water, chemicals, toxics, and children's health, among others. Check out EWG's website, www.ewg.org—it's a helpful resource that breaks down critical research in an easy-to-digest format and even identifies specific brands to seek out or avoid. To simplify the process even further, you can look for EWG VERIFIED™ personal care, cleaning, and baby products; you can find these product lists on EWG's website.

Understanding your own values will help you through your decision-making process as a conscious consumer. Using the Less but Better Method is one part of that process, and when it comes to supporting businesses and brands, you'll have to make those decisions based on your values and standards. These can be informed by your own research, as well as the research behind certifications like B Corporations and by organizations like Good On You and EWG, among others.

Before we wrap up this section, there's another element that's worth taking into consideration, which is whether to shop online or in person. While simple on the surface, each of these options comes with its own highly complex set of variables. Let's briefly break them down. You'll have to weigh the pros and cons yourself based on the choices you make to determine what might be more eco-friendly.

Online shopping: Online shopping can be more eco-friendly than driving from store to store, depending on a few considerations. For example, if you choose regular shipping over fast shipping, then fulfillment centers can efficiently pack and route their trucks, reducing the carbon footprint compared to driving yourself to various stores and back. Same-day or next-day shipping, on the other hand, can result in higher carbon emissions due to fewer packages being delivered by the same truck and less efficient delivery patterns. Another element to consider is that many online stores now are direct-to-consumer, meaning that they don't have to maintain storefronts that require energy and land for the in-person shopping experience. Ordering online, however, does result in excess packaging like bubble wrap, cardboard, and tape. Additionally, product returns from online purchases are sometimes tossed instead of resold because it's expensive for companies to assess the products to make sure they meet the standards to resell.[6]

Shopping locally: An important benefit of shopping locally is that it supports your local economy, particularly when businesses are locally owned. Shopping in town also means that you can plan your routes and shopping trips for efficiency, and you can opt for walking or local transit to reduce your footprint (if you live in town, that is). You can also bring your own bag when you go shopping and avoid the extra packaging waste associated with shipping. In-person shopping may also result in fewer returns, because you can see and perhaps try the product before taking it home.

The moral of the story is, don't fret too much about online versus local shopping, as the math can get messy and the decision-making fatigue around that calculation could have you giving up altogether. Again, this may come down to your personal preferences about where your values lie in terms of what sustainability means to you. However, try to avoid returning products bought online!

Each time we make a purchase, we are essentially voting with our dollars for the world we want to live in. That can be a lot of pressure if we analyze every single cent that leaves our wallet; I'm not suggesting being perfect, because it's nearly impossible to make sure every single cent we spend is making the world a better place. That said, I believe we still have opportunities to support the sharing economy, secondhand market, and conscious businesses, and that's what makes us conscious consumers.

Let's walk through an example of making conscious decisions as a consumer. Say your coffee maker just broke. What's the first thing most people do? Toss it and buy a new one without much thought to anything other than perhaps the price and the color. Instead, as a conscious consumer, the first thought should be to repair it if possible. If it's repairable, then great! If not, we can use the Less but Better Method to find a replacement, because it's safe to say that a coffee maker is now on your wish list. Let's go through the steps:

1. **Use what you have:** Do you have any other way to make coffee in your home? Maybe there's a French press collecting dust in the back of your cupboard. Do you actually need to buy a replacement for your coffee maker? If you can't find the French press, see the next option.

2. **Borrow it:** This doesn't seem so practical if you plan to use a coffee maker daily. But if your parents have a coffee maker they never use, perhaps it's worth asking if you can "borrow" it indefinitely. If not, head to the next option.

3. **Rent it:** Again, this one isn't exactly a practical option for a coffee maker requiring daily use. Head to the next option.

4. **Swap it:** Check your local Facebook swap groups or Bunz to see if you can trade something you no longer want for a coffee maker someone else is hoping to unload. No luck? Head to the next option.

5. **Find it for free:** Scout your Buy Nothing groups and other online marketplaces for a free coffee maker. If you can't find what you're looking for there, then head to the next option.

6. **Make it yourself:** Hmm, unless you're an inventor or an engineer, this one isn't likely applicable. Head to the next option.

7. **Thrift it:** Check out your local thrift stores or eBay and see if you can get a good secondhand coffee maker. If you can't find one that you love, then head to the next option.

8. **Buy it better:** If none of the above options worked out and you still haven't found a beloved new-to-you coffee maker, it might be time to buy one new. See if you can find a durable, high-quality coffee maker (read the reviews!) in a price range that works for your budget. Based on your personal values about what makes a brand's products better, set about finding the right coffee maker for you!

The Less but Better Method isn't intended to be a wild-goose chase to find perfection in the products we buy. Like most things in life, it won't always be possible to find things that check all of the boxes. Prioritize the criteria that are most important to you, from longevity and price to eco-friendliness. Keep it simple: Buy less and buy better.

Engaging in Accountability

While reading this chapter, you might have started to reflect on the brands you love that no longer align with some of the sustainability values you're developing. A common criticism when it comes to encouraging people to become conscious consumers is that it puts the onus on consumers to do the legwork. I agree that it shouldn't fall completely on your shoulders to undertake exhausting research to maintain a conscious lifestyle. There are many other players who have key roles in ensuring that all businesses work toward sustainability. These key players include business owners, government officials who develop legislation and policy, politicians, lobbyists, and so forth, and their efforts and accountability are integral to moving toward a more sustainable system as a whole.

Depending on how invested you are in pushing your beliefs forward, you might want to consider reaching out to corporations so that they change their business practices and becoming an engaged citizen with your government to speak up for these issues. We are more than just consumers; we are citizens. Let's build the future we want to live in! If you want to get really engaged, here are some ideas to get active:

Write letters and send emails: Let corporations know what types of changes you'd like to see that would support a more sustainable and ethical future. Get specific with your suggestions.

Create a petition: Have a particular issue that you want addressed? Start a campaign to tackle governmental issues or company policies, spread the word, and collect signatures in person or online using a platform like Change.org, which helps people and organizations start digital petitions for free.

Get involved: Volunteer, join committees, show up for public meetings, and become an engaged citizen. While it's great to share your concerns on social media, it can be empowering and impactful if you become involved directly with the causes you're passionate about.

Reflection and Checklist

What do you envision for the future? Not everyone is focused on making the world a better place through sustainable living. For now, it's people like us who, through many of the actions outlined in this book, are pushing the needle toward a more equitable system of production and consumption. But ultimately I hope that one day—sooner rather than later—*everyone* (environmentalists or not) will be able to support a better system without much thought. Am I a dreamer with my head in the clouds? Maybe. But I won't let that stop me, and I don't want that to stop you either!

As I mentioned before, this isn't about being perfect. It's not about having the most sustainable wardrobe (khakis and skater shoes or otherwise), or touting our electric vehicle, or shouting from the rooftops about our plant-based diet. Sure, we can do those things. Just beware that left unchecked, it can become another form of keeping up with the Joneses, except these Joneses are more eco-friendly. No one should ever have to feel bad if they can't afford a more sustainable option. There's no space for eco-shaming here. Being a conscious consumer isn't intended to be a new form of guilt. It's about learning more, doing better, and becoming an engaged citizen who fights for change with your dollars and your actions.

Are you ready to dip your toes in, or perhaps even dive into the deep end of becoming a more conscious consumer? Here's a checklist of activities from this chapter to get started:

- Get off the hedonic treadmill of consumption
- Reflect on this question: How much of what I have and what I want (in terms of both material things and lifestyle) is based on my own true desires?
- Consider how you can support a circular economy by having your products repaired and buying better-quality products
- Understand the true cost of each item you purchase
- Shop secondhand
- Follow the Less but Better Method for getting things on your wish list; the more you follow this framework, the more it will become second nature
- Develop your own set of values for what makes a brand *better* (i.e., sustainable and ethical) based on your preferences and beliefs
- Become an engaged citizen

Chapter 4

Pantry Goals

"No bag, please!" Overhaul your kitchen with zero-waste grocery shopping and a pantry audit

No More Chips

Even though it's been a few years since we started our low-waste journey, my husband still hides his Doritos and other packaged treats from me in strange places. In kitchen cupboards that we don't use to store food, or behind the passenger seat of his car—easy for him to reach but still well hidden from me, I guess? It's like if I don't see it, it never really happened. Regardless of where he hides them, I always manage to stumble upon his secret stash of cookies and chips. The odd thing is that I've never told him that he couldn't enjoy these packaged treats—I'm certainly not the packaging police! I simply care about our impact and wish to minimize it where possible.

Between food waste, excessive packaging, and cleaning supplies, oh my, the kitchen can be a trashy place. But it doesn't have to be that way. Caring about our trash shows that we care about ourselves, others, and the planet. Can you imagine achieving both peace of mind and sustainability in your kitchen (and your life)? Sounds like a pretty good deal, and it's totally attainable.

To keep things straightforward, I've broken down this chapter into five stages: conduct a pantry audit; find low-waste grocery and kitchen options in your community; learn how to grocery shop with your reusable bags and containers; reduce your food waste and start composting; and last but not least, a brief discussion on what to eat.

You might ask yourself, *Why bother going through this effort?* Well, there are loads of benefits. You'll likely save money (because you'll shop smarter and buy only what you need). There's a chance you'll eat healthier and maybe drop a few pounds you've been keen to lose (because lots of packaged foods tend to be overprocessed and high in sugar). Your cupboards, drawers, and shelves will become serene instead of chaotic (because you'll simplify your stuff, cut out much of the waste, and store your food in more aesthetically pleasing ways that will also make everything easier to find). You'll have to do this only once, and after that, you'll be set with a new system that takes less time out of your day than your previous routine. Then, when you're glowing as you cook and entertain, your friends and family will be asking WTF type of magic fairy dust you sprinkled in your kitchen to get such incredible results. You'll smile and tell them that being eco-friendly never looked or felt so great.

Oh, and just in case you live in a remote location far from people and things and stores, maybe on a mountain or on your own tiny island, let me just say that *results will vary:* Your low-waste kitchen will look different from mine and different from the next person's. It all depends on things like where you live, what resources are available, your waste-management system, your budget, and what you hope to achieve. Let's get to it!

Do a Pantry Audit and Start Creating Your Dream Kitchen

Making my kitchen less trashy was a *heck yes* moment! It totally transformed my kitchen into a blissful place where I could see all of our food and buy only what we needed. Let's rewind first: Before I reduced my waste in the kitchen, I had already been struggling with the mess of excess food packaging in my pantry. Trying to find the bag of almonds, or anything else, for that matter, was a bit like playing the lottery: I'd throw my hand into the pantry and hope for the best. Alas, I might pull out the dried cranberries or quinoa instead and then have to go fishing again. *Sigh.*

Not only that, but anything at the far back of our pantry was very likely to be a few years old. The funny thing was, though, that the ancient packaged food didn't look much different than it had the day I bought it. *Is it still good to eat?* I'd often wonder. The truth was, I knew I could still feed it to my husband, and he wouldn't be able to tell the difference, nor would he care. Then I'd feel too guilty and decide to avoid unknown diseases and throw it out. Better safe than sorry, right?

This system was totally trashy, because soft plastics aren't recycled where I live and because we were throwing out expired (i.e., *uneaten*) food. My first step was to clean out the kitchen. I pulled everything out of the fridge, cupboards, and drawers that was edible, then threw out anything that had gone bad and donated packaged foods that were unopened and not likely to get eaten (and that weren't expired, of course!). Foods that could be nicely rehomed into glass jars (like quinoa, almonds, and dried cranberries) were transferred and their original packages were thrown out. Now I could finally see what was in my pantry and fridge, no guessing necessary. Putting everything back into my pantry in glass jars or other beautiful containers was basically a dream. It was tidy AF. I felt very accomplished, and you will, too. It's that instant gratification we all need to keep going.

Keen to maintain momentum, the next step I took was to make a list in my notebook of every food item in the kitchen that I knew we'd buy again. As I worked on this, my husband was on the couch watching TV and looking at me like I was slightly crazy but also knowing I was onto something big. Then, beside each item, I scribbled down its type of packaging in these categories: compost, recycling, and/or trash. From there, I eagerly made a list of low-waste and plastic-free alternatives for each food item. If I couldn't find a low-waste option, we had to decide if we'd continue to buy the item or not. Doritos didn't make the cut, or so I thought. . . .

Now it's your turn to do a pantry audit. Take a look around your kitchen and find a place to start. You can make a whole day of this activity or do it in smaller spurts—whatever is convenient for you. Check every nook and cranny in your kitchen to make sure you have a full picture of all the waste generated there—and make sure you have a pen and paper close by when you do it. Go into each cupboard, look through all of your shelves, and don't forget your deep freezer or cold cellar, if relevant. Make a list of all your food items (or at least the important ones), identify which part of the waste stream its packaging fits into (based on your municipality's waste-management system), and then determine what low-waste and plastic-free alternatives (if any) are available instead. Sometimes the low-waste option may be to make it from scratch yourself, but not everyone has time for that. You do you. Remember: It might be a pain in the ass now to go through this process, but think of how your kitchen will look and feel when it's done. It'll be magical, and you'll eventually forget you ever put any effort into it.

Let's run through an example. If you found some boxed cereal in your pantry and know it's an important part of your life, check off the recycling symbol for the box and the garbage symbol for the bag (if that's how your municipality sorts these materials) and then list an alternative available in your community. Some substitutions may require a bit of research up front, but the next time you go to the store, you will know to shop for cereal with your own container or reusable bag to avoid packaging waste altogether. Or you might decide to give up cereal completely and go for low-waste granola (homemade or store bought) instead. Complete this task for everything in your kitchen that you'd like to swap for low-waste alternatives. To shorten this process, you can group together similar foods, such as spices, to keep things moving quickly.

Here's an excerpt of the pantry audit I undertook in my kitchen:

Food Item	Compost?	Recycle?	Landfill?	Low-Waste Option
Boxed cereal	NA	Box	Plastic bag	Buy cereal from the bulk section of the grocery store
Bread	NA	NA	Plastic bag	Buy it at the bakery in my own bag or in brown paper bag that can be composted or recycled
Condiments	NA	Plastic containers	NA	Choose condiments in glass containers instead of plastic
Hummus	NA	Plastic container	Plastic wrap-around container	Make it myself (actually my husband makes it most of the time; I'd better note that or face a tough conversation)

Food Item	Compost?	Recycle?	Landfill?	Low-Waste Option
Meat	Scraps	NA	Cling wrap, Styrofoam tray	Get meat wrapped in unlined butcher paper that can be composted
Milk in plastic bags	NA	NA	Plastic bags	Milk in returnable glass bottles
Peanut butter	NA	Plastic container	NA	Buy it in glass jar instead of plastic
Produce in plastic bags	NA	NA	Plastic bags	Buy naked produce and skip any in plastic bags
Yogurt	NA	Glass or plastic	For individual servings, the peel-off lid	Yogurt in returnable glass jars or glass containers that can be reused or recycled

This is also a good time to tidy up and organize your kitchen. As you do this, be sure to make it simple for yourself and other members of your household to sort waste properly. If applicable, the compost container and recycling bin should be easy to access, to avoid compost or recycling materials ending up in the trash bin. If you plan to start shopping with your own containers and reusable bags (more on that below), you can get a head start in the kitchen by transferring already-packaged items into see-through containers. When you can effortlessly see *and find* what foods you have in the kitchen (no fishing expeditions necessary), you'll be more likely to eat them and avoid food waste. Not only is this activity a great way to reduce your trash, but you'll also feel accomplished at the end, with an organized, beautiful, and tidy-AF kitchen that everyone will love. Remember that the kitchen is the heart of your home, so it's worth taking the time to make it beautiful and easier to navigate while reducing your waste.

Find Low-Waste Grocery and Kitchen Alternatives

While I made my list of low-waste food alternatives, some of the available options weren't immediately obvious. It took a bit of research and experimentation to figure out my new system. For starters, I wondered, where would I buy milk? In Ontario, milk is often sold in a carton or in plastic bags (yep, bagged milk is a thing). Hunting for a sustainable way to purchase milk became a fun opportunity to explore as I searched for alternatives to my go-to brands. I found I could buy milk in a glass jug and yogurt in a glass jar that could each be returned for a deposit—you may have noticed that listed in the chart above. For bread, I discovered a few bakeries in town where I could shop with my own reusable bags or leave with bread in a brown paper bag instead of plastic. As for eggs, I have a few friends who raise chickens, and I discovered that I could return egg cartons to them to be reused (this was also the case at the farmers' market). This adventure gave me the chance to get to know people at the farmers' market a little better, and to shop at smaller grocery stores rather than big-box chains. That said, we still shop at typical grocery stores, and we look for the low-waste and plastic-free options they have and—mostly—skip the rest.

When you're doing your pantry audit, you may not figure out all of the alternatives immediately (kudos to you if you fly through the process!). It can take time to find the right options and resources for you in your community, but it can be totally fun, too. Enjoy the experience; ask friends, family, and co-workers for recommendations. You may frequent old places in new ways or find hidden gems you never knew about. I personally loved this part because shopping at big-box grocery stores can feel impersonal by comparison. Although I'd visited smaller, locally owned shops and grocery stores before reducing my waste, I started frequenting them much more often. I enjoy seeing familiar faces. You'll likely experience something similar when you become a regular and get to know certain staff and they get to know your low-waste inclination. Other days, you'll get someone new and you'll need to explain your shopping preferences again—but that just helps spread the word about being less trashy.

To stay motivated, remind yourself of why you're embarking on this lifestyle change in the first place. Was it that documentary you watched, perhaps *A Plastic Ocean*? Or a book you read, maybe David Attenborough's *A Life on Our Planet*? If you feel like giving up because you're sick of explaining your shopping preferences to a new cashier for what feels like the hundredth time, consider rewatching that inspiring documentary or finding new ones with similar messages (like, a somewhat-earth-shattering-and-shit-has-hit-the-fan-but-we're-going-to-get-through-this kind of message).

This journey is about you and the changes you want to undertake. Celebrate your wins, big or small. I remember being so excited when I discovered I could buy yogurt in a glass jar that was returnable for reuse. When you remember why you started making changes, and you celebrate your small wins, you'll be even more pumped to take the next step in the process. Invite your family and friends to join, and it'll be even more enjoyable. You'll get in some extra bonding time *and* you might even convert others to be more sustainable, too (more on that in chapter 8). For now, take it moment by moment and day by day. Eventually what seems challenging now will become second nature down the road. The following is a list of easy tips to find low-waste options in your community. Go forth and conquer! Simplicity and low-waste living are right around the corner.

Tips for Finding Low-Waste Alternatives in Your Community

- If you have access to a dedicated zero-waste grocery store, then *score!* You'll likely find everything you need and more and be able to reduce your waste *a lot.*
- Shop the bulk section of your grocery store or find bulk grocery stores (if you can't shop with your own containers, using the plastic or paper bags they provide can be less wasteful than the packaging you'll find down regular grocery aisles).
- Bring your own containers shopping (more on how to do that below).
- Carry a reusable shopping bag, tote bag, or basket while shopping.
- Use reusable bulk bags / bread bags / produce bags.
- Find and use bottle returns (for items like milk, yogurt, nut milks, alcohol, etc.).
- Shop your local farmers' markets for fresh, local, and package-free foods. Farmers' markets are great for supporting low-waste shopping. You may find vendors happy to take back and reuse things like cardboard baskets (that peaches or apples sometimes come in) or those green baskets they use for berries, which are normally made of plastic or cardboard.
- Support your local bakeries for delicious breads and less-packaged treats.
- Make your own wine.
- Brew your own beer or refill a growler at a local brewery.
- Grow your own produce in a vegetable garden (in your backyard, on your balcony, with hydroponics, or in a community garden).
- Can or preserve foods.
- Buy larger quantities (bulk) with less packaging.

- Choose more easily recyclable materials (e.g., glass, paper, aluminum) if a low-waste option isn't available.
- Make foods from scratch that would otherwise come packaged (like my husband does with hummus).
- Consider not buying certain items at all if low-waste alternatives aren't available (like *Doritos*! well, to be honest, Hubby still buys these . . .).
- Join a local CSA (community-supported agriculture) in which you purchase a farm membership in advance of the season (providing financial stability to farmers) and get local, fresh, and in-season produce on a regular basis.
- If you're shopping online and having your groceries delivered, call the grocery store in advance to find out how waste can be reduced while their staff do your shopping (like avoiding produce wrapped in plastic and skipping plastic bags if at all possible).
- Seek out reusable-container programs. More and more programs are becoming available to facilitate deposit returns for things like food takeout containers, coffee cups, and grocery totes. One example of this is Loop, bringing brand names such as Häagen-Dazs and Seventh Generation to your home in reusable containers that are then returned for reuse (even the shipping package, the Loop Tote, is reusable).

A quick note on low-waste alternatives: You may not find everything you need in a less trashy form. Or maybe you just don't have the time. If you can't find a swap for a staple you don't want to stop eating, then don't. It's as simple as that. There are so many reasons you might not find an alternative, ranging from allergies (issues of cross-contamination in the bulk section) to accessibility (perhaps there's a need to buy precut, packaged produce) to availability of specialty and ethnic foods, plus issues like organic produce often being sold in plastic packaging so it doesn't get mixed with nonorganic items, etc. For example, my husband and I just couldn't get over the price of fancy cheeses. While it's nice to have fancy cheese from a specialty cheese shop (where we can shop with our own containers) once in a while, we just couldn't justify the price of this household staple on a regular basis. We opted out and still buy cheese in plastic on occasion. Additionally, I've been eating gluten free, and I can't get package-free bread that's also gluten free. Gluten-free bread comes wrapped in plastic to avoid contamination, so I'll treat myself. While I could try to make it from scratch, I'll just admit here and now that I don't enjoy cooking and baking; I only do it because I have to eat. C'est la vie.

The pantry audit is intended to be an empowering process. The very fact that you're making this effort to begin with is badass. Don't let the eco-shamers get to you (because they are out there, somewhere, trolling and ready to pounce on "flaws"). Eco-shamers could be people you know (maybe friends, perhaps family) telling you that what you're doing isn't enough or that it won't make a difference. They could also be people online, especially if you're sharing your journey on social media. Haters gonna hate. I've had my fair share of critics online, mostly through Instagram comments. Comments like "If you consume dairy you can't possibly be an environmentalist." Well, that *is not* their friggin' business—and I don't believe in an all-or-nothing approach. I had to take a moment, and still do, to reflect on negative comments like that one to determine how relevant they are to my life. Upon reflection, I decide whether or not to make additional changes in my life inspired by others' insights or to just let each comment go. Anyone who is blatantly mean, however, can see themselves to the door. What's the takeaway message here? Celebrate your victories and be okay with your imperfections (and don't even worry about labeling them as such; it just is what it is). Do what's best for you and your life, your home, your family, etc. It's no one else's business.

Low-Waste Kitchen Swaps

Going low waste in your kitchen with a pantry audit and changing the way you grocery shop to reduce waste is great. But why stop there? Food waste and food packaging aside, many of us also rewrap food in the kitchen using disposables. Plus, you might also be using disposable plates, cutlery, and cups for all of your fabulous dinner parties and backyard barbecues. Here is a full list of other (nonfood) swaps you can easily make in your kitchen. This is also a chance to get resourceful. If it's not in your budget to buy beeswax wraps, then could you try making them at home? There may be workshops in your community or online to help you transition to some of the kitchen swaps below. It's a great opportunity to meet like-minded people, too.

- Aluminum foil → silicone baking mats, reusable tinfoil
- Cupcake liners → silicone cupcake liners
- Dish soap in a bottle → solid dish soap bar, dish soap refills
- Disposable coffee cups → reusable travel coffee cups
- Disposable coffee filters → reusable coffee filters, French press
- Disposable plates, cups, and cutlery → reusable plates, cups, and cutlery (don't have enough? borrow from your local tool library or from family and friends!)
- Disposable synthetic washcloths → reusable washcloths, compostable dish cloths
- Disposable water bottles → reusable water bottle, tap water filter

(if you don't like the taste of your tap water, or if you're dealing with possible contamination issues, consider researching clean and low-waste drinking water options)

- Paper lunch bags → reusable lunch bags
- Paper napkins → reusable cloth napkins
- Paper towels → cloth towels, UNpaper Towels
- Parchment paper → silicone baking mats, reusable tin foil
- Plastic bread bags → reusable cloth bread bags
- Plastic dish scrubber → bamboo scrub brush with natural bristles
- Plastic grocery bags → reusable shopping tote bags or bins
- Plastic produce bags → reusable cloth produce bags
- Plastic sandwich bags → reusable silicone bags, storage containers with lids
- Plastic straws → skip the straw, reusable straws
- Plastic wrap → wax wraps (such as beeswax wraps or soy wax wraps), cloth bowl covers, storage containers with lids
- Tea bags → loose-leaf tea with an infuser

Remember that you won't make all of these substitutions overnight. Be prepared to slowly transition your kitchen. You may want to make *all the swaps* immediately (you go-getter, you!), but that can be overwhelming. Make a priority list of the transitions you'd like to undertake and go from there. If you're wondering what to do with your disposables once you start reusing, you can either use them up, give them away to family/friends, or donate unopened ones to charities that will gladly take them. Another option is to repurpose; find other household duties that your disposables could fulfill. For example, plastic sandwich bags are still great for corralling your stuff (perhaps safety pins or hair clips), so this is your chance to discover something in your home that needs to be organized and use the plastic bags in new ways. You could also use some of your disposables as art and craft supplies, either for yourself or for your kids.

How to Shop Low Waste

As a society we've become accustomed to all of the packaging that comes with grocery shopping. It's *normal* to buy our fruits and vegetables in plastic containers and cardboard boxes. Even if produce is unpackaged at the store, people commonly grab disposable plastic bags to corral all of their loose fruits and veggies. (I've been there, done that. Won't do it again.) Once at home, those plastic containers, bags, and boxes go in the trash and recycling bins. Tons of trash can be brought home from fruits and veggies alone, yet this is only *one section* of the grocery store. You already know how much trash grocery shopping creates, because you did your pantry audit, right? Yep. It's trashy.

Beyond that, plastic bags abound in the bakery to package breads and other baked goods. Fresh-baked cookies and treats come in plastic clamshell containers, too. In the middle aisles of the grocery store, you'll find more hard and soft plastics, cardboard, and some Frankenstein concoctions of plastic/metal/cardboard (e.g., juice boxes). Everything in every aisle seems to be overpackaged. Although it is necessary to have some packaging to keep food fresh, much of it is overdone. (Once you've read this book, you'll never look at packaging the same way again. You're going to see trash *everywhere!* #sorrynotsorry)

Yikes! Is anyone else exhausted by all the trash we deal with daily? Think about all the time and energy it takes to package, unpackage, and dispose of these materials. The thing is, we don't buy these products for the packaging. We don't go to the grocery store because we want plastic bags, cartons, containers, and in general, trash. We simply want the fruits, the vegetables, the food and beverages. Which is why it's time to change how we grocery shop and skip the packaging altogether when possible. This can be done in a number of ways, including by using your own reusable bags and containers. It's time to beat (and change) the system. With more and more people choosing unpackaged groceries, grocers will start to get the picture.

I have to admit that I was lucky when I first started my low-waste journey, because I had a bulk store nearby that had a reusable-container program. I could comfortably walk into that store knowing that I wouldn't be *that weird person* shopping with my own reusable containers and bags (while it's not weird, it can feel weird). You don't always know if the request to use your own containers and bags is going to be well received. The best you can do is ask!

Let's talk about the tools you'll need, like reusable bags. Reusable bulk bags and produce bags come in all shapes, sizes, and materials. If you don't already have reusable bags, you can make them or buy them. Lightweight reusable bags are perfect for both produce and bulk dry goods. They come in a variety of sizes and should be light enough that you don't have to worry about their weight—just shop as you normally would with disposable plastic produce bags. How many will you need? That depends on how much you buy each time you go shopping. Start off with a set of five reusable produce bags (I prefer see-through mesh for these) and another five reusable bulk bags if you plan to use those for your dry goods (I prefer solid cotton for these) in small and large sizes. One of the best parts of shopping with your reusables is that you'll likely inspire others to do the same. I've had plenty of positive conversations with other customers who have a lightbulb moment when they see me shopping this way.

If you have heavier reusable bags for bulk shopping, you may want to consider getting the tare (i.e., weight) of the bag when it's empty. This way, the cashier can subtract the weight of the bag when you go to pay. Be sure to check what the policies are for the store you're shopping in—they may want to weigh your bags and containers before you start shopping, or they may not participate in this type of shopping whatsoever. Chances are, you'll have the most success using reusable containers and produce and bulk bags at dedicated zero-waste grocery stores.

You can also shop with your own containers; again, this will be best received at zero-waste grocery stores. There are plenty of beautiful images all over social media of pretty jars filled with delicious bulk foods. Jars are a great plastic-free option for containers, but keep in mind that you can use any type of container you have at home, no fancy jars required. Your jars can be brand-new, secondhand, or reused from previously consumed products (e.g., jam and sauce jars, well washed, of course). If a container will serve the purpose you need it to, then why not use it for your shopping? To avoid any messes with pouring bulk food into a small container with a big scoop, use a canning funnel. These are often supplied at zero-waste grocery stores, because they don't want messes either!

As anyone will tell you, there is a learning curve involved here. Zero-waste-dedicated grocery stores will make it super easy, if you have access to them in your community. If not, it's about finding ways to shop as low waste as possible with whatever is available to you. The more people who request this type of package-free shopping, the more it will become normalized and available. Go forth and fill your jars and bags! Here's a quick list of how to do it:

- Make your grocery shopping list.
- Choose appropriate containers and reusable bags for the items on your list; be sure to have a canning funnel on hand.
- Get the tare of your container (either at home, if you can, or at the shop; staff may be available to help you with this).

- Mark the tare weight on your container (a chinagraph pencil works great for this purpose).
- If you're sewing-savvy, you can sew the tare weight of your reusable bags right onto the bags (or find someone else to do it).
- Once you're at the store, fill your containers and note the item number of each food item (you can use your cellphone and take photos of the number to reduce the waste associated with writing the number on a tag).
- Pay! The cashier should subtract the tare weight of your container. Feel free to ask how this works in case you want to double-check.
- Share your zero-waste grocery haul on Instagram and tag @zero .waste.collective and #zerowastecollective! I'd love to see your progress!

A Day in the Life of My Low-Waste Grocery Shopping

Okay, now you've got a pretty good idea of how to go low-waste grocery shopping. Want to see it in action? Here's a breakdown of my low-waste shopping trips (check out Instagram while you're at it; there's loads of inspo). Plus, here's a quick tip for your grocery list: Keep one. It's that simple. My husband and I have a chalkboard where we keep a running grocery list, but a note on your phone works well, too. When we leave the house, we can snap a photo and know that we won't forget something. Here's our shopping system:

- Check the grocery list to determine what types of bottles, containers, and bags we need to shop with, then prepare the shopping kit.
- Make a list of the stores we need to visit for each item on the list.

- Plan out the trip and mode of transportation (walking, biking, or driving).
- Then out the door we go! Here are typical items we buy:

    Produce: Eating our fruits and veggies is a priority. We avoid plastic and packaging most of the time, which does limit our selection. For example, we can get fruits such as strawberries, blueberries, blackberries, grapes, etc. plastic free only when they are in season at the farmers' markets or by picking our own at nearby farms. I freeze local berries and sliced peaches to eat over winter (it's best to wash then freeze them on a cookie sheet before transferring them to a container—glass or silicone works well—to avoid your fruits freezing together). To corral produce, we use our own reusable mesh produce bags.

    Dry foods, baking ingredients, spices, etc.: We have a zero-waste bulk food store close by, and we bring our own containers and bags to fill up. The list of foods we can buy in our own containers is quite comprehensive, including peanut butter, coconut oil, and tea.

    Bread and baked goods: We have a lot of options for baked goods: the farmers' market, bakery at the grocery store, or local bakery. Or we can make baked goods ourselves at home.

    Canned foods: While it would be lovely to stay at home, Martha Stewart style, and make all of our sauces, condiments, and soups from scratch, *not everyone has time for that!* Maybe some people have time for that or are interested in making these staple foods from scratch, but it's just not happening in our household (although I do occasionally make soup, and it's *the best*). We still tend to buy canned goods from a variety of grocery stores in town, depending on where we're shopping that day. We look for BPA-free cans when possible (cans are lined with plastic to reduce corrosion). What I like about cans is that they are highly recyclable compared to Frankenstein materials.

Condiments and sauces: We buy these in glass jars and bottles or cans to avoid plastic. The great thing about glass jars and bottles is that we can wash and reuse them (I'm not a jar minimalist by any means).

Meat: We purchase meat at a poultry shop or local butcher instead of the grocery store to reduce packaging waste like Styrofoam trays. After watching a number of environmental documentaries (like *Cowspiracy*) and food documentaries (like *Food, Inc.*) we have modified our household diet to be healthier and less processed and to include far less meat.

Cheese: We occasionally buy fancy cheese from specialty cheese shops in our own container, but it's expensive. We still buy a big brick of cheese from Costco (we used to shop at Costco *a lot*, but since reducing our waste we find most options at Costco are overpackaged and not low waste at all).

Milk, yogurt, and butter: We buy milk in returnable glass jugs and yogurt in returnable glass jars. Both have a deposit system to make sure they get returned to the store. These are available at specialty food stores in town. We buy butter as is, wrapped in foil.

Eggs: We have friends with chickens, so we often get eggs from them. Once the egg carton is empty, we save it for our friends to reuse or drop it off at the farmers' market.

Snacks: We either make snacks from scratch (recipes on Pinterest have become my go-to) or nibble on things like nuts, seeds, fruits, olives, and cheeses. An incredibly important household snack staple is popcorn. We can get the kernels in bulk in our own containers, so we eat lots of it. Then, of course, there are moments when my husband brings chips or cookies home and hides them. Oh, and there was that one time I couldn't *refuse* buying Girl Guide cookies because, I mean, who can say no in that situation? I'm not a dream crusher.

A lot of the products we now regularly buy came with some trial and error at first as we explored new brands and options. There were things we didn't like, and once we used them up, we moved on to try something else. When we first bought milk in a glass bottle, we didn't like the taste. We found another brand with a bottle-return program, and that's what we still buy today. It also took some time to find out where we could get better prices for the same products. The same will likely happen for you. Try a bit of this, check out a bit of that, then find what you love the most and you're set. If you don't find alternatives you love, then you can decide to go without, make something yourself, or even arrange for a trade with someone you know. For example, if you have a friend who loves making tomato sauce from scratch, offer to trade some other food or a service you can offer for the sauce. Start getting creative! I have a friend who loves growing veggies and preserving, so we came up with a trade: I give her advice on using Instagram and she gives me garden-fresh veggies. Win-win. If you still can't find a good alternative after investigating those options, you might decide to stick to your original preference.

Reduce Food Waste and Start Composting

Admittedly, my household is still working to get better at reducing our food waste. We're not the only ones, as about one-third of food produced globally is wasted. Since becoming less trashy, we've been better at grocery shopping and sticking (mostly) to our grocery list, which helps us avoid buying too much food. Here are some tips to avoid food waste:

Keep that shopping list handy. Get out of the habit of picking up whatever piques your interest at the grocery store, *especially* if you're not sure that it will get eaten before it expires.

Plan meals. This one is super important. Planning meals will help you develop your shopping list and ensure that you have the food you need for each meal and snack. Make a master list of meals, then plan out week by week which ones you'll make, and you're set. If meal planning isn't your jam, consider finding low-waste meal-kit services (like Crisper, based in Toronto, Canada) delivering recipes with ingredients in reusable and returnable containers.

Do weekly meal prep. Meal planning makes meal prepping easier. Based on your schedule, choose a day that you can prepare your food for the week. Chop your fruits and veggies, cook the basics for the week (e.g., rice, eggs), batch a few meals, and then eat them throughout the week. Not only will you have delicious meals to eat every day, but having partially prepped lunches and dinners will help you avoid all of the packaging associated with eating out or ordering in.

Eat the leftovers. This. One. Is. Vital. Do your best not to let your leftovers go to waste. Eat them the following day or incorporate them into a new meal. Enough said.

Freeze extras. Made too much? Freeze it! You'll be thankful on a night when you don't feel like cooking and you can just pull out a premade, homemade frozen meal.

Store food properly. Some food should be stored in the fridge, and some food shouldn't go in the fridge. For example, tomatoes do best on the counter and avocados should go in the fridge only once they are ripe. Storing your food better will ensure that it doesn't spoil unnecessarily quickly.

Make sure your food is easily visible. This is especially important for food that is closer to expiration. You can keep a basket on your counter or in your fridge labeled "Eat me first." Any food that is hidden is likely to not get eaten and to meet its demise in your compost bin (because you will start composting if you don't already!).

How to Store Food Without Plastic

If you're looking to lessen your dependence on plastic in the kitchen, you may be wondering how you can store your food without it. Personally, I grew up with plastic *everything*. Tupperware and disposable plastic sandwich bags were the norm in my house. These days, with studies showing that plastics can leach chemicals into our food, people are looking to ditch plastic for nontoxic alternatives. Below I offer a list of alternative food-storage options for you to consider.

Beeswax (or Soy Wax) Wraps

- Perfect for keeping bread fresh; I like to use Goldilocks beeswax wraps. Wrap your bread and place it in a cloth bread bag and leave it on the counter to keep it as good as day one (for a few days, not weeks). Pop it in the freezer for a week, too (any longer and you run the risk of freezer burn).
- Wonderful for produce. If you have some sliced-up peppers or half an avocado, wrap it up in beeswax wrap, pop it in the fridge, and it'll stay fresh.
- Great for wrapping snacks or sandwiches on the go.
- Good for wrapping up cheese.
- Useful as a bowl cover; I like cotton bowl covers (waxed and unwaxed) from Your Green Kitchen.

Glass Jars and Containers

- Perfect for all leftovers. Use big containers for large portions and small containers for individual-sized portions (*hint, hint,* meal prep!); Glasslock containers are my favorite for this purpose.
- Wonderful for chopped veggies and fruits. Veggies such as carrots, celery, and radishes, among others, do well submerged in water.
- Great for storing meals in the freezer.
- Ideal for liquids like soup. Soup can be frozen in glass containers; just make sure you leave extra space at the top for expansion. Plus, make sure your soup isn't hot when putting it in the freezer (close to room temperature is best).
- Ideal for storing frozen fruits and veggies.

Reusable Silicone Bags

- Great for snacks and sandwiches on the go; I use both Zip Top containers and Stasher bags. I love using these for hiking and traveling because silicone bags are much lighter than other containers.
- Perfect for freezing all sorts of foods, including leftovers, meats, fruits, and veggies.

If you can rock reusables in your kitchen, you'll give your food a longer life before it ends up in your tummy. Good food-storage options will help your food avoid the trash bin, and any food waste that slips through the cracks can go into your compost bin (more on that below). Better yet, all of the storage options listed above can be aesthetically pleasing, so a beautiful kitchen is an unexpected bonus. Keep in mind that your jars don't even have to match—an eclectic collection of jam and sauce jars will be stunning compared to disposable plastic baggies. Ridding your kitchen of plastic will help you have a healthier kitchen, lessen your packaging waste, reduce your food waste, and improve your overall aesthetic. The benefits are undeniable.

Composting

Composting is an integral component of a low-waste kitchen. Food waste should never, *ever* go to a landfill. What happens when it does? Food waste that decomposes in landfills releases methane, a greenhouse gas that is at least twenty-eight times more potent than carbon dioxide. I will quickly admit that I have it pretty easy when it comes to composting, because this service is available through my municipality's waste-management collection. Before I went low waste, I was already composting. Even if you don't have access to municipal composting, it's not too much work to transition to compost at home.

What does *compost* even mean? Basically, composting is a process of letting organic matter (like food) break down naturally into a nutrient-rich fertilizer that can be used in your garden or added to your indoor plant pots (other ways to use your compost are listed below). It's nature's way of recycling.

What can go into a compost bin? It depends on your local guidelines. To give you an idea of how you might sort some of your waste, here's a list of things that I'm able to put in my compost bin—just remember this list will vary from region to region.

- Coffee filters/grounds
- Dairy products
- Eggshells
- Food scraps
- Fruits and vegetables
- Grain products
- Houseplants and flowers
- Leaf and yard waste
- Meat and fish
- Nuts, legumes, and seeds
- Oils (solidified)
- Paper soiled with food
- Paper towels and tissues
- Pet feces and kitty litter
- Pet waste
- Sawdust (no wood pieces or pressure-treated wood)
- Spices
- Tea bags

That's a pretty solid list! If you have a local waste-management system that accepts compost, be sure to find out what you can put into it. Something worth mentioning here is that bioplastics (such as compostable coffee pods) *are not* currently listed as accepted items where I live. Compostable plastics need to be processed in special facilities to break down properly, so before you start dropping bioplastics in your compost bin like the greenest mofo out there, make sure they are accepted. Take care to sort properly. Done and done.

If you don't have a municipal compost system, there are always other options. Here are some ideas:

- Compost in your backyard.
- If you're living in an apartment, there are plenty of alternatives, including worm bins and bokashi composting. A worm composting bin, also known as a vermicomposter, is a relatively inexpensive option for composting at home. Essentially, the worms live in a bin and will eat your food scraps (like fruit peels, but not meat or dairy), which get mixed with shredded paper to create compost. A bokashi bin will accept an even more comprehensive list of kitchen scraps (like meat and dairy) to break down with an inoculated bran in an anaerobic (i.e., free of oxygen) process that is worm free.
- Find out if any nearby farms or farmers' markets will accept your food scraps or compost.
- Join a community garden where you can contribute your organic waste to their compost bins.
- Freeze your compostable scraps until you go to the market and drop off your compostables at a community compost bin if available.

Whether you live in a single-family detached house or a high-rise apartment, there are great home composting options available.

What to Eat

It's not anyone's business what you eat. That said, it's still worth including in a book about sustainability a conversation about what a sustainable diet can look like. Let me clear the air here: I'm not a vegan (diet that contains no animal products), or even a vegetarian (diet that contains no meat but may contain animal products like dairy and eggs) for that matter. While I'm not a fan of labels, I am a stereotypical millennial in the sense that I have tried different—and even trendy—diets. As I write this section of the book, I'm currently gluten free (GF) and dairy free (DF), and I don't eat beef, but I doubt I'll maintain all of these specific restrictions forever. The reason I cut these things out of my diet is to address a potential health issue. I'm happy to take out the red meat anyway for environmental reasons, as it's one of the most notorious foods contributing to greenhouse gases. In this tiny yet mighty section of this book, we'll discuss what to eat and address the importance of organic food in the context of being less trashy.

I may not be a vegan, but just as the tiny mason jar represents one extreme end of the zero-waste-lifestyle movement, I believe it's possible to have a more sustainable diet without being perfect. Personally, I like to follow Michael Pollan's rules for healthy eating: "Eat food, not too much, mostly plants."[1] I'd say most of that is self-explanatory, but what Pollan means when he says "eat food" is to eat *real* food, not processed food that comes packaged and has an eternal shelf life. Not only is it healthier to eat more plants (and fewer animal products), but it's also better for the environment. In Jonathan Safran Foer's book *We Are the Weather*, Foer states simply that choosing to eat fewer animal products is probably the most important action an individual can take to reverse global warming.[2] It's certainly a much more inviting position than telling readers to go vegan *or else*. So if you take away anything from this section, it should be this: When possible, eat more plants and fewer animal products. If you're already on this train, kudos to you!

A Note About Organic Food

Organic farming is considered more sustainable for the natural environment because it doesn't use the highly toxic chemical pesticides common in conventional agriculture that can pollute the land and water. Also, high levels of exposure to these pesticides can have health consequences for farmworkers. Organic farming, by contrast, also promotes more biodiversity. It would seem, without going into academic-level research, that organic agriculture is healthier for people and the planet. This is a very simplistic overview of why organic food is better, but I think it gets the point across.

The problem I've run into, and you might too, is that organic produce often comes in more plastic packaging than its nonorganic counterparts at most grocery stores. One reason behind this paradoxical and frustrating reality is simply the need to tell organic produce apart from nonorganic, in part because organic tends to cost more and grocers don't want them mixed up. This brings us to a trashy dilemma if you buy organic: Do you choose nonorganic but package free, or packaged organic produce? Tough call. There are a lot of factors to weigh here, because organic farming is known to be better for the soil, plants, animals, and humans. But plastic packaging, of course, sucks on so many levels—plastic production is toxic, plastic packaging is often hard to recycle, if it's recycled at all, and what's worse is when it ends up in the natural environment harming wildlife (just to mention a few).

To measure what's better from a consumer perspective isn't easy. Just because produce isn't labeled organic doesn't mean the farmer wasn't working the land in a more sustainable fashion; in some cases, it may mean they just didn't get organic certification. It's nearly impossible for us consumers to tell the difference. The best way is to talk to farmers directly, at places like farmers' markets. When that's not an option, you'll have to prioritize for yourself what's most important to you at the grocery store, among things like your budget, your desire to buy low waste, and your preference to eat organic. The choice is yours.

Reflection and Checklist

Your efforts to clean up your kitchen routine may be the most frequently evolving portion of your journey to becoming less trashy, and your habits will likely become fine-tuned over the course of the year. Your preferences or diet could change; you may refine your shopping list or adjust your budget.

When I was just starting out, there wasn't anything efficient about my grocery shopping. The exploratory phase was exactly that: an exploration of the low-waste options available. Other times it was a matter of going all over town to find one specific food item that we loved (like package-free bread). It won't be like this forever, I promise. You'll eventually find the products you like, work out a grocery budget that makes sense for this lifestyle (overall we find it more expensive to grocery shop low waste, but we are eating better and more locally, and we believe in voting with our dollars), and know exactly where you need to shop to meet your grocery needs. You'll have a new and improved shopping routine down by the end of your Don't Be Trashy Challenge!

Perhaps you might stick to some of your existing purchases and incorporate a few low-waste alternatives, or you may go completely zero waste in your kitchen. How far you take the ideas in this chapter is up to you. With your new insights, you can find resourceful ways to reduce your waste. Ultimately, with this process you'll get a stream-lined, stress-free, low-waste, and aesthetically pleasing kitchen—*hello, magic fairy dust!*

Are you ready to tackle waste and bring simplicity to your kitchen? While a month for this project may seem too long for some people and too short for others, it's a great starting point. Here's a checklist of activities for you to tackle waste in your kitchen (feel free to add or subtract from this list based on the tips provided in this chapter):

- Do a pantry audit
- Find low-waste options in your community
- Create a wish list of low-waste kitchen swaps
- Go low-waste grocery shopping
- Develop ways you can reduce food waste
- Store your food with reusables
- Learn to compost (if you don't already)
- Remember your reusable bags!
- Eat more plants
- Support sustainable farming with your food purchases when possible

Chapter 5

All Things Bathroom and Cleaning

Say goodbye to a bursting toiletries bag and all the plastic bottles

Nontoxic Living

Early in the 2010s I started to discover that there are unsafe chemicals in many commonly purchased beauty products and cleaning supplies. It shouldn't have been that much of a surprise to me, especially when it came to the cleaning products. Many of them *do* have warning labels (hello, Captain Obvious). Despite this, it was easy for me to ignore the labels because I was brought up using these cleaners. But when it came to makeup and skin-care products, that's where the real shock set in. I had always trusted (i.e., assumed) that what was on the shelf for sale was vetted and safe to use. Loaded with new knowledge after doing a bit of research, I decided it was time to overhaul my beauty routine and scrutinize each of my skin-care and makeup products—and my use of hazardous cleaners, too. This revelation happened *before* my journey toward minimalism and low-waste living, but in retrospect, it's clear how interconnected these ideas really are. In this chapter we'll explore all things bathroom and cleaning with health, minimalism, and waste reduction in mind.

Not only was I concerned about the products I put on my skin, but I also wanted to make sure that the products we used to clean our home weren't adding to the toxic load our bodies were taking on. It makes me cringe now to think that I spent years cleaning with harsh chemicals without wearing rubber gloves. I wasn't protecting myself against the toxins that can penetrate the biggest organ we've got: our skin. Additionally, regularly breathing in toxic cleaning products can have negative impacts for our lungs.[1] This new awareness had me changing my habits pretty quickly, and it became my obsession to find and use nontoxic products.

What I didn't realize back then was that these efforts toward a cleaner and more minimalist life were laying the foundation for sustainable and low-waste living. All of these ideas are in alignment. Less stuff equals less waste. Fewer toxins equals healthier people and planet. These lifestyle changes were eco-friendly without my even thinking about it at that time, since the chemicals we use at home often find their way into the natural environment. Fast-forward to today, and I'm seeking out products that are healthier (i.e., nontoxic), with a minimalist approach (i.e., fewer products, more functionality), and that are low waste (i.e., have less packaging). My transition took years, but it doesn't have to be that way. I wrote this book so that you can use my learnings to help speed up the process for you.

In this chapter, you'll get to clean up your act in addition to helping the planet! Here's the game plan: get informed about ingredients; clean up your beauty routine; find ways to reduce waste in your bathroom; and clear out the cleaning supplies. This process may help you feel good on the inside and out!

Get Informed About Ingredients

Let's start off with some science, shall we? Don't worry, it'll be brief—chemistry is not my specialty, but I do believe it's important to have some understanding of what chemicals to avoid in our cosmetics and cleaning products. When I first began making product swaps, I started in the bathroom. The first few nontoxic swaps I made were shampoo, conditioner, body wash, and lotion. After doing some Google research and reading a few books from the library, I went in search of anything that was free of parabens.

You'll often find labels on bottles noting that a product is "paraben free." And with good reason; parabens are a class of commonly used preservatives known to disrupt hormone function.[2] Preservatives are important in cosmetics to reduce the growth of harmful bacteria and mold, which extends the shelf life of products. The problem with the hormone disruption is that it can negatively impact fertility, reproductive organs, and birth outcomes and increase the risk of cancer. Based on this research and information, I decided it was best to avoid products with parabens.

My favorite online resource to find out about safe versus harmful ingredients is the Environmental Working Group (EWG). In addition to parabens, EWG recommends avoiding products that contain the following ingredients, due to their toxicity and harmful characteristics:[3]

- boric acid and sodium borate (found in some diaper creams)
- butylated hydroxyanisole (found in food, food packaging, and personal-care products)
- coal-tar hair dyes and other coal-tar ingredients, including amino-phenol, diaminobenzene, and phenylenediamine (found in dandruff and psoriasis shampoos)

- formaldehyde (preservative in cosmetics)
- fragrances (can contain hormone disruptors and are among the top five allergens in the world)
- oxybenzone (found in sunscreen)
- parabens (widely used in cosmetics)
- PEGs and ceteareth (found in personal-care products and cosmetics)
- petroleum distillates (petroleum-extracted cosmetics ingredients)
- phthalates (often included in fragrance, as well as in some nail polishes, plastics, paints, and air fresheners, among other things)
- triclosan and triclocarban (antimicrobial ingredients in a range of personal-care products, including toothpaste and deodorant)

Be sure to check out EWG's website (updated regularly) for more details and to look at reviews of specific products you might have at home.

Armed with this knowledge, you can assess the products you have, check their ingredient lists, and make an informed decision about which products you want to keep using and which ones you want to eliminate from your household.

Did you know that the air inside our homes is two to five times more polluted than the air outside?[4] This is due to things like building materials, home furnishings, cleaning products, and air fresheners releasing harsh chemicals into the air, including ones linked to asthma, developmental harm, and cancer.[5] Here are a few great tips from the EWG website regarding toxic ingredients found in some household cleaners:[6]

- Avoid cleaners that include ammonia or chlorine bleach, because they are highly toxic, and when accidentally mixed together, they create a dangerous chloramine gas.

- Skip ingredients that contain hydrochloric acid, phosphoric acid, sodium or potassium hydroxide, or ethanolamines, because they can cause skin burns, blindness, and lung irritation.
- Avoid air fresheners and scented products that don't disclose their fragrance ingredients, as these products may trigger allergies and could contain endocrine (hormone) disruptors.
- Skip triclosan, which is an antimicrobial that is linked to increased allergen sensitivity and may disrupt thyroid function.
- Avoid quaternary ammonium compounds, which are chemicals associated with asthma, reduced fertility, and birth defects in animals.

Government regulation of these types of ingredients will vary depending on where you live. The European Union, for example, has banned many ingredients from personal-care products that you'll still find in use in U.S. products. Sadly, that puts the responsibility on us consumers to make sure what we're using is safe. While that's not ideal, there are online resources, like EWG, that can help us make informed decisions about which products to use at home and on our skin. Knowing all of this, are you ready to clean up your beauty routine?

Clean Up Your Beauty Routine

When I was in my early teens, I adored those multitiered makeup kits that included what seemed like hundreds of eye shadows, dozens of lip shades, and a variety of concealer/blush/bronzer colors. It's not that I wore makeup regularly, but I had a lot of fun playing dress-up. The options and combinations were endless. Surprisingly, I didn't grow up to be glam in the beauty department; I'm pretty minimal in terms of my aesthetic, but I still like a good eye shadow and mascara—and some foundation to conceal my rosacea, a red and bumpy skin condition on my face.

Even though my beauty routine was pretty simple, back before I overhauled it, I still managed to accumulate plenty of makeup, lotions, and potions. I seemed to have bought in to all the bells and whistles, everything from creams and deodorants to shaving paraphernalia, makeup, and various shampoos. Sound familiar? When I first started minimizing my toiletries, I reduced the number of nail polish bottles in my largish collection down to a handful. Once I came to terms with the fact that I didn't paint my nails often enough to justify having much of a collection at all, I decided to give away *all* of my nail polish to create more space and simplicity in my life. This transitional approach made it easier to see that minimizing made sense for me. Had I gotten rid of all my nail polishes at once, I might have quickly regretted it and relapsed, perhaps rebuying some of the items I had decluttered, which would have been counterproductive.

Our daily beauty routines are significant, even if we don't give them much thought. Most people tend to have one. Some might simply take a shower, brush their teeth, and maybe put on some deodorant. Others might put on some makeup and body lotion as well. Some might have extensive evening routines. How we start the day can impact how we feel for the rest of it. If getting ready in the morning is a bit chaotic, the fact that you often find yourself spending your precious minutes digging through a bowl of hair elastics, clips, and costume jewelry attempting just to fish out the bobby pins probably isn't helping. Especially when what you really need to be doing is walking the dog, packing lunches, or actually eating breakfast. One easy way to take control of your routine is by simplifying it. I'm going to walk you through a four-step process to detoxify, simplify, and streamline your beauty routine so that it'll be the perfect way to start and end your day, every day.

1. **Review your current routine.** First things first. Make a mental note as you go through your day of what your daily beauty routine is in the morning, during the day, and at night. Ask yourself the following questions about your toiletries and beauty products:

   - What's my current flow (i.e., what do I do each morning, and in what order do I use my products)?
   - How many products do I use each day? What are these products?
   - What products am I using the most? (These are the ones that you replace often.)
   - Which products do I barely touch? (These ones are probably collecting dust or are still mostly full, even though you've had them for a while.)
   - What might I have too much of? (These might be your bigger collections.)
   - Are all my products effective (i.e., doing what I want them to do)?
   - What products do I find essential? (If you were stranded on an island and could bring only a few things with you, what would they be?)

   Through this reflection exercise, establish what's essential to your routine and what's unnecessary. While you're at it, let your mind wander as you consider what a simplified routine each day could look like for you. This exercise will help you minimize in the steps that follow.

2. **Filter out your products by their ingredients.** Pull everything out of your drawers, shelves, cupboards, etc. Wherever you keep your makeup and toiletries, find them and corral them into one

space. Lay everything out on a towel to avoid anything leaking onto the surface you're working on. Then, using informational resources like the EWG website, assess each of the products you use and decide whether or not you will keep it based on your knowledge of the ingredients. Discard anything that no longer meets your standards. While you're at it, recycle or throw away anything that's expired. Most beauty-product containers are hard to recycle, so check out what local options might be available to you. I suggest finding out if you have access to a TerraCycle (a company that can recycle almost any form of waste) bin in your community; L'Occitane, for example, has a partnership with TerraCycle to recycle beauty-product containers. Before you move to quickly replace discarded products, be sure to go through the following steps.

3. **Slowly minimize.** Now that you've detoxed your beauty products, take a good look at what you have left. Recall that I didn't get rid of all my nail polish in one go. Now it's your turn to incrementally minimize the products in your daily routine to transition toward less. Try not to toss everything all at once (unless you know *for sure* that you won't need certain items), because you may regret some of your decisions. Instead, as you slowly downsize, you can decide what is actually important to you to keep versus what should really be removed. Before you opt to throw away or recycle your unwanted (and presumably open) beauty products, find out if you can rehome them with family and friends. Unopened and unexpired products may be accepted as donations at local nonprofit organizations or shelters as well.

4. **Start fresh with your new and improved routine.** Once you

become comfortable with your reductions, step into your new and simplified routine. Keep in mind that this won't happen overnight; it may take a few weeks or even months to create your ideal setup. The goal is to have a nontoxic beauty routine with only your favorite and regularly used toiletries and makeup. This is the perfect time to consciously fill in any product gaps you have, using your knowledge from chapter 3 ("Conscious Consumption"), researching ingredients using resources like the EWG website, and equipped with tips from the next section on ways to reduce waste in the bathroom.

Some people have become convinced that they need a ten-step beauty routine. Take one of my book editors as an example. Donna shared with me that she had bought in to the idea that she should be using toner and primer. She was using *a lot* of products, including toner, serum, sunscreen, primer, foundation, setting spray, and finishing powder. She bought in to the marketing (as we all have) that using more products would be better, but after deciding to streamline her life, she simply uses one serum, plus a sunscreen/foundation hybrid (both in glass bottles) each morning. Her skin feels just as good, yet with a simpler routine. Now she has more time *and* money, although she still loves lipstick and keeps a small collection, and that's okay! Like Donna, I've reached a point where I know exactly what makeup products I want in my life, making it quick, easy, and inexpensive to get ready in the morning or for a fun evening out. Here's a list of the essentials I use to get ready for my day:

- Shampoo and conditioner bar
- Bar soap
- Body lotion
- Face wash
- Face moisturizer

- Deodorant
- Toothbrush
- Toothpaste
- Floss
- Hairbrush
- Concealer
- Foundation
- Eye shadow
- Mascara
- Lip balm
- Lipstick (occasionally)

When you develop a solid routine, you'll be more likely to stick to your tried-and-true heavy hitters. It saves money and avoids waste. It's as simple as that!

Find Ways to Reduce Waste in Your Bathroom

I can admit that finding perfectly nontoxic and low-waste makeup products is a bit like looking for a needle in a haystack; there are so many options popping up these days! Getting the right skin cream for your dry skin or a foundation in the closest color to your skin tone is more challenging than it should be. The point is, reducing your waste should *simplify* your life. More and more toiletries and beauty products for sustainable and low-waste living are becoming available, but it always takes some trial and error to find exactly what you need. Keep that in mind, because buying new and exciting "zero-waste" products can result in more waste if the process of trial and error you go through results in a pile of unwanted products that weren't the right color, texture, scent, or whatever. I confess that I passed along plenty of unwanted products to my friends and family as I tried new things. Before you dive into the world of these new and clean products, my biggest piece of advice is this: Use up what you already have first, then find replacements when needed. After that, it's time to put your new knowledge and skills to work by using the Less but Better Method for adding new products to your bathroom!

In the pages that follow, I do my best to ensure that your phase of exploration isn't as meandering or long-lasting as mine was. Some of the products mentioned are my favorites, and many are popular swaps in the low-waste community. I know that many of these products won't be readily available to everyone at a nearby store, which means there might be some online shopping involved (a topic covered in chapter 3). It can be quite an onerous task to assess every single purchase we make based on what's more sustainable and ethical (taking everything into consideration, like labor, manufacturing, materials, toxicity, shipping, packaging, etc.), but I believe it's better for people and the planet to buy products made with those values. If that means shopping online and having products delivered to your door, then I believe it is a better choice than buying products from the nearby shop that are less transparent.

The following list includes products that will work for some people and not others. While I do mention some brands below, keep in mind that there are so many more options available and new ones that become available every day. Lately I've noticed there have been more eco-friendly and low-waste product options entering the market, which is amazing! This is what happens when we vote with our dollars: More products that align with the values people are demanding become available. It's not magic; it's simply supply and demand. With that in mind, here's a list of low-waste swaps you can make in your bathroom:

Toothbrush: Plastic toothbrushes have a bad rap for being a source of plastic that's filling our oceans and our landfills. That's a fair point, so what are the alternatives? Bamboo toothbrushes, like those from Brush Naked, are one option, with a backyard-compostable handle (or find out if your municipality accepts these in their compost system, which mine does not) and nylon bristles that need to be removed and thrown out. Other options include reusable handles with interchange-able toothbrush heads, like those made by Yaweco, or brands like Pre-serve, with toothbrushes made from 100 percent recycled plastic, supporting a circular economy. I'm still trying to decide which option is my personal favorite! Regardless of which option you or I choose, I do find it's hard to avoid packaging, although most come in cardboard. While the options aren't perfect, we're heading in the right direction!

Toothpaste: There are lots of options available, including tooth pow-der. This is used much like normal toothpaste; you can simply wet your toothbrush, dip it in the powder, and brush. Toothpaste in jars and tooth tabs, which you can crush in your mouth before you start brush-ing with a wet toothbrush, are also available. There are great low-waste toothpaste options in tablet form from brands like Nelson Naturals and Lush, but I admit that I still use mainstream toothpaste from time to time. Not into the whole toothpaste-in-a-jar thing? Check out Davids natural toothpaste in a metal tube. Do what works best for you! And perhaps talk to your dentist, too.

Mouthwash: There are a few low-waste options for mouthwash, like getting it in a glass jar (Lucky Teeth has a few options), finding refill stations locally and at zero-waste-dedicated shops, or getting concen-trated tablets that you drop into water (like those from Georganics and Bite).

Floss: Instead of using plastic, some people opt for silk floss in a reus-able glass jar or stainless-steel container, but it's not always a vegan-friendly option if that's a requirement for you. In my experience, silk floss is more likely to tear and get stuck in my teeth because it's not as strong as the synthetic version, so be mindful of that if the space between your teeth is tight. KMH Touches makes floss (made from either silk or corn husk fiber) called Flosspot and is well known for its zero-waste floss, as is Dental Lace. If you're looking for a plastic-free, vegan, and refillable option, try bamboo-fiber floss (available from brands like Lucky Teeth). You can also get a water flosser (like Waterpik), which is endlessly reusable. There are lots of options, so choose what works best for you and your preferences!

Razor: This is one of my favorite low-waste swaps. I previously used a razor with a reusable plastic handle and disposable blades. It was gender-specific and expensive, so much so that I'd often buy the blades marketed to men because they cost less. When I began to reduce my waste, I switched to a metal safety razor by Albatross Designs and I'll never look back. I even find my metal safety razor offers a closer shave, too! I save my dull stainless-steel blades and bring them to the scrap metal recycling facility in town. The blades themselves are super inexpensive, and the box I purchased has enough for the next few years. Just be sure to find out how to recycle the blades properly where you live. While buying a safety razor may be an investment up front (anywhere between fifteen and fifty dollars), it's worth it in the long run. Plus, you might even be able to find one secondhand! There are some pretty cool-looking vintage options out there. Curious how to use one? Here are some tips:

- Get to know it! Feel its weight and shape, and put the razor blade in carefully (always holding the shorter edges that aren't sharp). How the blade is inserted will vary depending on your razor.

- Lather up! I simply use a soap bar, but there are shaving-specific bar soaps, too. Want an extra lather? Use a shaving brush.
- Safety razors need to be held at a thirty-to-forty-degree angle to shave. There's no need to apply pressure; just let the weight of the razor do the work. You'll need to move with your curves to maintain the angle, which will take some practice, so go at your own pace. It took me a few months to get comfortable with using a safety razor.
- This is a close shave, and your skin may be a bit dry afterward, so moisturize!
- How often you replace your blade depends on personal preference, but they typically last about ten to fifteen shaves. When it's time to swap, dispose of your blades according to your municipality's system or use safety razor blade take-back programs like the one available through Albatross Designs.

Not sure if you're ready for a safety razor? You may want to check out alternatives like Leaf Shave, a reusable razor that also uses safety razor blades but handles more like a disposable razor—it's similar in shape and pivots as you shave.

Toilet Paper: One option for eliminating toilet paper from your life completely is to use family cloth, which is exactly what it sounds like. Using reusable cloth to wipe your bits makes for a lot less trash going down the toilet. The idea is that each family member has their own set of cloths to use for number one and number two. You need a sealed laundry bin for used cloths and must wash these cloths separately from other laundry, using hot water. I've never done it, but there are some good YouTube videos if you want to learn more. I have, however, used my Kula Cloth (available online and at select stores), which is an anti-microbial pee cloth perfect for outdoor activities like hiking and camping and for travel. It's the best way to leave no trace (i.e., toilet paper) in the wilderness. Alternatively, you can opt for a bidet (which can come stand-alone or as a toilet attachment like Tushy), which allows you to wash yourself after going to the bathroom. Bidets are known for being more hygienic than toilet paper! That's because, according to advocates, using toilet paper just wipes poop around, whereas a bidet will actually wash it off. Another option is to choose toilet paper made from recycled materials that comes in plastic-free packaging, which you can buy online from brands like Who Gives a Crap and Reel Paper or at local shops.

Shampoo and conditioner: I've tried the refill option a few times at my local eco-friendly shops but ultimately landed on shampoo and conditioner bars by Unwrapped Life as my go-to. Like many other brands, the company makes shampoo and conditioner bars for all hair types that can be bought online or in select stores. This is an easy swap you can make as soon as you run out of the shampoo and conditioner you're using right now. If you want to go the refillable route, see if there's a local shop that will let you bring your own bottle to fill up on shampoo or conditioner when yours is empty (go back to chapter 4 for details on shopping with reusable containers). Alternatively, brands like Plaine Products make it easy to refill via online orders and shipping. With each new shipment you receive from Plaine Products, you'll ship back your empty Plaine Products stainless-steel containers for reuse.

Body wash and hand soap: I use bar soap for both of these (typically from the Soap Works, made in Toronto). Finding naked bars of soap for all skin types is easier than ever! You can look into liquid refill options similar to those described above for shampoo and conditioner. Another cool option that's available for hand soap is buying highly concentrated mix (like tablets from Blueland or powder from EarthSuds) that you combine with water in your own reusable hand-soap bottle.

Body and face lotion: One of my favorite solid lotion bars is made from beeswax (I get mine from Bee Savvy); it's smooth, smells nice, and is perfect for travel. Although most of the time I get a glass jar refilled with body lotion locally (same concept as described above for refilling shampoo and conditioner). Looking for some DIYs? Check out the blog at www.thezerowastecollective.com/blog! As for face lotion, I try to find products that come in glass jars or metal containers that can easily be washed out and recycled. Dew Mighty, for example, has a concentrated face serum that comes packaged in cardboard, and you can reuse the tin you buy with your first order.

Deodorant: There are a lot of deodorant options these days, but I often use Meow Meow Tweet, which you can buy in either a jar or a solid form that is packaged in a cardboard tube. Other low-waste brands that come to mind include Native, Routine, Schmidt's, and Ethique, among so many others. There are also refill brands like Myro that offer a reusable tube that can be refilled. Simple as that!

Menstrual products: This transition was a game changer for me. I swapped out disposable pads and panty liners for their reusable cotton counterparts (brands like Öko Créations and Aisle—or you can sew your own, following online tutorials) and leak-proof panties (like those from Knix and Thinx—I use both). I simply rinse my reusables in my washing machine on a quick rinse cycle, then add in my towels or sheets that I'd be washing anyway. You can hang these to dry or dry them in the dryer with towels and sheets. I swapped out tampons for a menstrual cup (brands include DivaCup and OrganiCup). Now I don't spend any money on disposable menstrual products, saving me thousands of bucks in the long run. It grosses some people out, but whether we use reusables or disposables, we still have to deal with a bloody mess. Might as well do it in a way that creates less waste and saves money!

Makeup wipes, cotton balls/rounds: Go for the reusable options! You can buy or sew (using an online tutorial) specialty makeup wipes and cotton rounds, or just use a facecloth.

Cotton swabs: There are both plastic-free and reusable alternatives to cotton swabs. I've recently switched to a reusable version made by LastObject, which comes in its own carrying case, making it easy to travel with. It's simple in that you use it, wash it, and reuse it again! LastObject makes reusable beauty-specific swabs, too.

Hairbrush: If you need a new one, opt for natural and renewable materials like wood and natural rubber that are sustainably sourced; I personally use Tek brushes.

Tissues: We swapped these out for reusable and washable cotton hankies. We simply use them, toss them in the laundry, then wash them with our towels. We do still keep a box of disposable tissues for guests.

Makeup: More and more options are becoming available for low-waste, cruelty-free (i.e., no harm to animals), and nontoxic makeup, such as Elate Cosmetics, Kjaer Weis, and Cheekbone Beauty (and so many more!). What should you look for in low-waste cosmetics? Opt for minimal and recyclable packaging when possible. Refills are another great option; instead of buying a new compact each time you buy a powder foundation, for example, find brands that let you buy a refill for your existing compact. Most of these cutting-edge brands are available online, some are available at mainstream shops like Sephora, and while more limited, there are refillable makeup options at some zero-waste and eco-friendly shops!

Lip balm: Instead of buying the plastic tubes I used to get, I now buy lip balm in a reusable tin or compostable cardboard tube. Making lip balm is another good DIY project—find a recipe online and whip up a batch!

While this list is fairly long, it's not as long as it could be and is meant only to cover the basics. These days there's a product out there for any and all of our needs, even the needs we didn't know we had. My biggest tip to you, when it comes to making swaps, is to avoid shiny-object syndrome. Just because there are a bunch of cool and pretty zero-waste swaps out there, that doesn't mean you need them. Over time your needs will become clear, but as with any other type of consumption, it's easy to get drawn in by new products. And it might be even easier to pull your wallet out in the name of the environment, right? If you buy *this* product, you'll save the planet! Do your best not to get sucked into that mindset, and always stick to what you really need.

Clear Out the Cleaning Supplies

Is it just me, or is there a bottle of cleaner for every surface in our homes? For some reason we need a tile cleaner, which is different from a tub cleaner, which is different from a kitchen cleaner, which is different from a floor cleaner, which is different from a stainless-steel cleaner. I get it, some materials do require extra care, but when you stop to think about it, the wide array of specifically tailored cleaners can seem a little excessive. And a lot of them are pretty harsh, too. There might be times when these strong-AF cleaning products are required. Most of the time, though, we don't need to use harsh chemicals in our homes.

I was introduced to more eco-friendly cleaning supplies as I transitioned to a low-tox life, and it wasn't until I started making this transition that I realized I had way too many cleaning supplies to begin with, when a few would have sufficed. Even though I had been decluttering for years, I had never made the effort to sift through our cleaning products. They just sat there, some hardly used, if ever. Maybe you can relate? Each time a new magical cleaning product comes along, do you jump on the bandwagon? Well, my friends, that willy-nilly approach ends today. It's time to detox your cleaning supplies, figure out what you really *need* to clean your home, and streamline your cleaning routine with the low-waste options that work best for you.

1. **Detox your cleaning supplies:** Similar to the way you decluttered your cosmetics and toiletries, pull out all of your cleaning supplies, regardless of where they live in your home. Then, using informational resources like the EWG website, assess each of the products you use and decide whether or not you will keep it based on your knowledge of the ingredients. Discard anything

that no longer meets your standards. You can give these away to friends and family (informing them of your new learnings, of course, so they can also make informed decisions); donate un-used products to organizations that accept unopened cleaning supplies; and throw away anything that remains, based on your municipality's requirements (if any of your cleaning supplies are hazardous, your municipality may have special drop-off loca-tions for those products). For example, aerosol cans in my city are considered household hazardous waste and must be dropped off instead of placed in my curbside trash bin. Plenty of clean-ing supplies come in aerosol cans, and many of us wouldn't have given much thought to the fact that these products might be considered "household hazardous waste," right?

2. **Simplify your cleaning routine:** After going through the exercise above, you may wind up with fewer cleaning supplies. It's worth going through your remaining products to assess what's left. Ask yourself the following questions:

   - Will I use everything here? (If not, create a game plan to give away or donate what you won't use.)
   - Am I missing anything that I need to clean my home? (The following section has plenty of suggestions for low-waste cleaning swaps!)

   While it's freeing to simplify and declutter, it's also less wasteful to use up what you already have at home.

3. **Low-waste cleaning options:** Once you're ready to fill in any gaps in your leftover cleaning supplies, consider the following low-waste options. Again, similar to the transition of your beauty products, this process won't likely happen quickly. The goal is to have a nontoxic, minimalist, and low-waste cleaning routine, so

that you're left with more space, more money, and perhaps more sanity as cleaning your home becomes a breeze! This is the perfect time to consciously fill in any product gaps you have by using the Less but Better Method (see chapter 3), researching ingredients from resources like the EWG website, and implementing some tips from the low-waste cleaning swaps here:

- **Product refills:** Product refills are great if you have access to zero- and low-waste shopping options where you live, or you can order these online! For example, the company Pure has cleaning-product refill locations across Canada, and Common Good has refill stations across the United States (with a few in Canada, too). Bring your own containers and fill up what you need in store. If you don't have access to that type of shopping, you can also do low-waste bulk refills at home with companies like the Bare Home and Common Good. While these options do not eliminate all waste, they are still more eco-friendly compared to their conventional counterparts.

- **Concentrated cleaners:** Concentrated cleaning supplies are another great way to reduce waste at home. A well-known version of this is Dr. Bronner's castile soap, which is a highly concentrated cleaning solution that, when mixed with water, can be used to make various cleaning products: all-purpose cleaner, veggie and fruit wash, toilet bowl cleaner, and body wash, among so many other uses. Other low-waste, concentrated cleaning brands like Blueland and Dropps have concentrated tablets to reduce plastic waste at home, or you can try Tru Earth laundry strips. Another bonus of concentrated products is that they weigh less for shipping, thereby reducing their carbon footprint.

- **Solid bars:** Similar to concentrated cleaners, solid bars reduce waste by eliminating single-use plastic. No Tox Life carries the Dish Block, which is a concentrated cleaning bar that's great for washing dishes as well as cleaning other areas of your home. Multi-use products like this help cut waste, save money, and simplify cleaning routines.

- **Make yourself:** My all-time favorite, easy-to-make all-purpose cleaner is simple! Which is why I love it. All you need is a spray bottle (you probably have a bunch of these that you can repurpose), water, vinegar, and (optional) essential oil. Here's the breakdown:
  - 1 part vinegar to 2 parts water (e.g., 1 cup of vinegar, 2 cups of water)
  - 10–20 drops of your favorite essential oil per 3 cups (I like to use a citrus mix)

  This all-purpose cleaner is safe for most surfaces, but you should avoid using it to clean granite, marble, and soapstone countertops and solid-wood furniture, just to be on the safe side. Vinegar is a great disinfectant, with many antibacterial properties; that said, it does not kill viruses like COVID-19.[7] If you're keen to find out which cleaners are ideal for cleaning to protect against the spread of COVID-19, be sure to check your government website for recommendations.

- **Cleaning accessories:** What about the stuff we clean with? Well, there are low-waste and eco-friendly options here, too. Skip the paper towel and use a reusable rag or cloth for cleaning. Or opt for a Ten and Co. Swedish sponge cloth, which can hold up to ¾ cup of water, can last up to a year (depending on how much use it gets), is easily cleaned in the dishwasher or washing machine, and can be composted at the end of its life. Other options include cleaning instruments like dish scrubbers and brooms that are made from durable materials like wood (compostable), metal (recyclable), and natural fibers (compostable).

Reflection and Checklist

Your body is a temple. This is a good reminder to take damn good care of your body, because it's precious. We may have been in the dark for a long time about the products we've been using, but we can make informed choices. It's better for both people and the planet.

This month has combined a few different topics. It's been about using nontoxic products in your home, from toiletries to cleaning supplies, and we've also tackled ways for you to reduce waste in your bathroom and in your beauty routine. To top it off, hopefully you've felt inspired to minimize your possessions in these categories. Now is a good time to reduce the number of new items (say, another lipstick or specialty cleaner) that might find their way into your collections. Simplifying your beauty routine and cleaning supplies will help you detox your life, streamline your home, save money, and benefit the planet.

The lesson here is to scrutinize any new products that are vying to enter your home. If it's a priority to you, try to find eco-friendly *and* nontoxic products. And to reduce clutter, remember to find products that serve multiple functions. If it's a single-tasker, ask yourself if you really need it or if another multitasking product could get the job done. This month is about going back to basics. Here's a quick checklist of this month's activities:

- Get informed about ingredients to reduce the toxicity of products you use
- Detox your cosmetics and toiletries, then minimize them based on your preferences
- Create a new beauty routine that is nontoxic, simplified, streamlined, low waste, and cost-effective
- Detox your cleaning supplies, then minimize them based on your preferences
- Create a new cleaning routine that is nontoxic, simplified, stream-lined, low waste, and cost-effective
- Try making your own all-purpose cleaner, and DIY your heart out if that's your thing
- Ditch paper towels and plastic-based cleaning supplies for durable, low-waste, and eco-friendly options

Chapter 6

Outfit Repeater

*Clear out your closet, develop your personal style, and
break up with fast fashion*

Fashion, More Than Just a Fling

Hi, I'm Tara, and I used to be addicted to fast fashion. It's been a few years
now, and it's taken some work, but I've slowly weaned myself off my need to
be perpetually buying cheap new clothes. And you can, too.

I don't think anyone can argue with the kick-ass confidence that comes
along with wearing a great outfit. But there's a highly superficial and trendy
aesthetic saturating social media. There are many influencers out there frol-
icking around in a brand-new outfit *each day* for their followers to admire
and purchase. You won't see them wearing the same outfit twice, and that's
exactly the lifestyle they are promoting: the constant consumption of fashion, among other things. Fast fashion is like dating on Tinder: The romance
is fast and passionate, then rapidly ends up in the trash. The boredom comes
quickly, and you're back online looking for something new. That is not my
thing. I have a small wardrobe (even if it didn't start that way, as I shared my
story of closet woes in chapter 2) and I love it. Sure, I occasionally add new

pieces. However, it takes time to make a purchase because I want to fall in love! I'm not interested in a short-term fling, you know? I want to be in it for the long haul.

Unlike traditional fashion, which had two to four seasons per year, fast fashion has fifty-two-plus seasons, with brand-new styles coming out weekly. That's why it's called *fast* fashion. The styles often knock off high-end fashion, making the affordable pieces even more appealing to the masses. Who wouldn't want to get a thousand-dollar runway look for twenty-five dollars? Sadly, that cheap price comes with high environmental and human costs. It's time to move beyond the one-night stands of fashion.

The problem with fast fashion, in a nutshell, is this: The manufacturing system and supply chain source cheap labor with low labor standards; they manufacture in countries with low environmental standards; and they generate tons of textile waste from unsold products and from products that are tossed in the trash by consumers when the next fashion trend comes out. This probably isn't news to you, as thankfully this information has been widely covered in books, documentaries, and the news. If we know the issues, then why are we still buying in to fast fashion? Ask yourself that question if you're still addicted to buying cheap clothes on the regular. After reflecting on your answer, ask yourself: Am I ready to make a change?

Where does our shopping addiction come from? There are no doubt a variety of influences, including things like needing to fill a void in life, putting a bandage on a personal insecurity, and simple boredom. For example, if you're seeking external validation, you might work on your wardrobe as a way to make yourself appealing to others. Perhaps you want people to envy you, to think you're cool, fashionable, and trendy. At the end of the day, these are superficial reasons for filling your closet and draining your bank account. While there's nothing wrong with enjoying fashion and wanting to look nice or perhaps glamorous, it just doesn't seem healthy to me to be using fashion as a crutch for other issues in our lives. I know what it's like! I've been there.

During my undergraduate studies, I went on a semester abroad to Poland. Coming from my small town of Guelph, Ontario (remember, *not* New York!), I found the city of Kraków way trendier than I'd anticipated, and the clothes I had packed and carried across the Atlantic didn't fit the bill (except for my hot-pink tube top). I quickly made an effort to go shopping to fit right in with the European vibe. Given my student-sized (i.e., nonexistent) budget, fast fashion fit the bill. Did I know about the issues of fast fashion at that time? Yes. Were they important to me in that moment? No. My main goal was to fit in. Kraków might not be known as a fashion capital of the world, but let me say, people were well dressed, and they looked great! I wanted to look and feel great, too.

I imagine that what I experienced in Kraków is completely relatable. We all want to look and feel great. That's, in part, why we buy fast fashion. It's like getting a touch of luxury that many of us otherwise couldn't afford. To exacerbate the problem, we now have social media profiles with images of ourselves and our belongings that can become instantaneously available to our friends and followers. We get a hit of happy chemicals each time we get a new like or comment, so we can become addicted to sharing more content.[1] To share more content, we need more fashion, and to get more fashion on a tight budget, we need to keep shopping for fast fashion. It's a vicious cycle, and it's resulting in a lot of textile waste as we move from one trend to the next. North Americans send twelve million tons of clothing to landfills every year, 95 percent of which could have been reused or recycled.[2] This is the cycle that needs to change in order to transition to a more sustainable lifestyle. This raises the question: Can we still get creative with fashion without always falling back on fast fashion? Of course we can!

Instead of being focused on all the latest trends, I now put my attention on the styles that appeal to me the most and aim for timeless instead of trendy fashion. I've slowly developed a personal style that I'd describe as "relaxed chic"—relaxed because my lifestyle is pretty low-key, but chic because my taste is slightly elevated and stylish. Whether or not anyone else would agree is not my concern, and it shouldn't be yours either. Building a (mostly) sustainable and ethical wardrobe that works for me means that I don't need to listen to the peanut gallery. I'm a proud #OutfitRepeater, too! A sustainable wardrobe is a rewearable one. If you think it's uncool to wear the same outfit twice, consider this: "In 1930, the average American woman owned nine outfits. Today, that figure is 30 outfits—one for every day of the month."[3] I have a feeling that you own more than nine outfits! I know I do.

If you're ready to make this change, then buckle up! Here's what you'll do: assess your closet and develop your personal style; find ways to fill your closet sustainably; and become an outfit repeater. Listen, I promise that this is worth your time! Have you ever looked at your closetful of clothes, thinking that you have nothing to wear? Even if you are one of those people buying sixty new pieces of clothing per year? That's because we shop *mindlessly;* we don't stop to consider if the clothes we're buying really work for our lifestyles, whether the pieces feel comfortable, or even if they look that great on us. You know you've been there, done that. Who hasn't?

But those days are over, my friend, because now you're going to clear the clutter, define your style, and buy only what fits the parameters you develop for yourself this month. You're going to feel and look great after this process. The funny thing is, you'll probably get a lot of "You look fantastic!" and "What's your secret?" comments *anyway,* because you'll be killing it in the style department after this.

Assess Your Closet and Develop Your Personal Style

Even if your high school style doesn't resonate with you now (I'm not planning to dye my hair pink again anytime soon!), there's a good chance that the overall style you like today has some longevity in your life. Think of it as the foundation of your aesthetic, even if a few things change along the way, such as your preferred length of dresses and skirts or the cut of your shirts. As for me, I love denim. I've loved it since I was a child (thanks, Mom!), and I expect to love denim for the rest of my life. And I don't mean just a little bit of denim. I'm talking about a denim-on-denim situation, like a full-on Canadian tuxedo. So look at your closet and figure out what you really love, need, and wear, and let go of the rest. Perhaps you decluttered some of your clothes, shoes, and accessories in chapter 2. This time around you'll take an even closer look at your wardrobe, which may help you pare down even further. The questions below are tailored to help you do so. From there, you'll have a fresh start to develop your own style!

Assess and Streamline Your Closet

This section is intended to complement the decluttering guide from chapter 2, so depending on how far you got in decluttering your home, you may have already tackled your closet quite thoroughly. If not, then let's dig in. To prep for decluttering, pull out all of your clothes and accessories, then make your sorting piles: keep, sell, donate, trash. Just a heads-up: You might not need to trash anything at all if your local donation center will take worn clothing for textile recycling. Once you have your piles ready, pull out all your clothes from all your closets, grab your shoes, jackets, accessories, everything, and start sorting! To make it a little less overwhelming, you can do one category at a time. For example, do all clothing at once, followed by shoes, followed by jewelry and accessories. Ask yourself these questions for every single item to help you decide which pile an item belongs in:

- Do I *love* it?
- Does it fit well?
- Is it flattering?
- Do I wear it often?
- Do I remember when I wore it last?
- Do I know if I'll wear it again soon?

If you answered "no" to any of these questions, then consider that item to be on the chopping block; it might be time to let it go.

A simple rule of thumb is to get rid of anything that doesn't fit (pregnancy excluded). While you may have dreams of fitting into something, the best thing to do is to keep and wear clothes that fit right now. If you're planning to lose or gain weight, you can choose to keep those items with a specific timeline in mind. Ultimately you will choose what's right for you, but hear me out: I know I've held on to some clothes way longer than necessary because I envisioned losing that extra five pounds. I didn't take that goal too seriously, so the clothes just sat around unused. I ended up donating the items that didn't fit, and I actually felt better for it. You know why? Because every time I saw this one pair of shorts, it made me think that I needed to lose weight. That wasn't a healthy mentality. My recommendation: Don't keep clothes around if they just add to your feelings of guilt and shame. Nobody needs that!

Be ruthless in your closet culling, but if you have difficulty deciding on a few pieces, you can give it a bit more time. A word of caution, though:

Don't give yourself too much time, because you might just land back at square one. Another way to help yourself discover what to get rid of is to assess your lifestyle and how you spend your time. For example, if you split your time working in an office, going to the gym, and lounging at home, with the occasional formal event thrown in there, then think about how many clothes you need for each of those activities based on how much time you allocate to each. You'll need office clothes, gym clothes, lounge-at-home clothes, and a formal outfit or two. Plus PJs if you use them. Maybe you're a nudist, I don't know! If you attend formal events only, say, three times per year, then it doesn't make sense to have a closetful of formal wear. Make sure your clothes reflect how you actually spend your time. If you live on a farm, then you don't need clothes for city life.

Develop Your Personal Style and Conscious Closet

Before I pared down, I didn't take much time to consider how each new piece of clothing would fit into my wardrobe or overall look. I'd buy a skirt because I liked the color and pattern, only to get home and find that I had nothing to wear with it. The automatic solution would be to get a new top; mission accomplished, right? Not so much. Instead, building a conscious closet means being more intentional about adding new pieces to our wardrobe, and to do that, we need a solid foundation. That foundation is our personal style. Developing your style is important because it'll be your guide to building a closet that you'll actually wear, that reflects your real-life activities and personal preferences, and help you become more intentional with new additions to your closet. Whether you want to dress as glam as Beyoncé or low-key like Angelina Jolie, it's time to be your own stylist and get to work!

Developing your style is a combination of function and beauty. Here are some tips to help you refine your style blueprint:

Know your lifestyle. Develop your wardrobe around your activities and lifestyle. Make a list of all the events and functions you need clothes for and what types of clothes you need for those activities. Also make note of how often you need those types of clothes. For example, if you go to the gym, how often do you do that? Three times per week? More? Less?

Dress for the weather. If the winters where you live are cold and long, then be sure to take your seasonal wardrobe into account. Determine how much of each weather type you typically encounter. For example, if you live in an area where it seems to rain perpetually, you'll probably want more variety of rainwear than someone who lives in a dry, sunny place.

Choose a color palette. Doing your laundry is probably one of the best ways to find out what you wear the most. Next time you do a load of laundry, check out what colors you're washing, because those are the ones you're evidently wearing the most. Use this as the foundation of your color palette, and add other colors you love, that suit you, and that you know you'll enjoy wearing.

Create a vision board. I use Pinterest for this, and I have a board called "My Style: Relaxed Chic Vibe" where I save outfit ideas that I love the most. You can create an IRL (in real life) pin board and save magazine images of styles you like, too. Whatever works for you, start collecting outfit ideas and styles you think you'd love wearing.

Copy outfit ideas. Using your vision board, start copying outfit ideas to get a sense of what looks and feels good on you. Mix and match clothes from your wardrobe to re-create your favorite looks.

Try things on. Go on a "shopping" spree, but don't actually buy anything. Using the outfit ideas you've gathered, go to a store and try on different combinations to figure out what works for you and what doesn't, and make a list of your favorite styles, outfits, and looks. This is a great way to discover if the looks you like in photos make you look and feel good in real life!

Create a uniform. To keep things simple, you might want to create your own uniform. Steve Jobs, cofounder of Apple, was well known for always wearing jeans, a black mock turtleneck, and sneakers. Maybe you don't want to wear the exact same type of outfit day in and day out, but you could create your own outfit formula and mix and match from there. Keeping it simple will help you avoid stockpiling too many clothes and will help you get ready quickly every morning.

Make lists. Lists are a great way to help you stick to your style. List the materials you like wearing (e.g., organic cotton, linen, Tencel); the silhouettes you love (e.g., pencil skirts, culottes, A-line); the vibe you're going for (e.g., organic, punk rock, athletic, sexy); plus any other lists that will define your style best.

Create a title and mission statement for your style. I decided my style title is "Relaxed Chic." My style mission is to be comfortable yet stylish and to have the clothes that I need for the various activities that I participate in, including working from home, doing yoga, going hiking and camping, participating in networking events and conferences, and going to the occasional gala or other formal event like a wedding. I live in a climate with four distinct seasons and need attire that can handle the heat and humidity of summer and cold and wet winters, plus everything in between. My go-to material is denim, and my color palette is mostly black, gray, navy, pink, and coral. Based on all of the information you've collected for yourself above, develop your own title and mission statement!

Once you've defined your style, you're good to go! It may take some time to tweak, and you'll likely modify it as your life changes. Whether you have kids or pick up new hobbies, life will change your style accordingly. Through this process you may have identified some gaps you need to fill. I always keep a wish list on my phone of the pieces I'd like to add to my closet. Right now, one key piece I'd love to add to my wardrobe is a pair of denim overalls (skinny-jean style). Once you've decluttered your closet and defined your style, you might want to make your own wish list, too. I find that having a wish list, much like writing a grocery list, will help you avoid impulse shopping. When you avoid impulse shopping, you'll save money and avoid adding to the landfill with unnecessary textile waste. Everybody wins! Next we'll talk about how to fill those gaps a bit more sustainably!

Find Ways to Fill Your Closet Sustainably

Back in high school when I read Naomi Klein's *No Logo,* the problem was clear, but no alternatives really existed at that time. At least, none were available where I lived, nor appealing, if we're being frank about it. Sustainable clothing has been notorious for being disappointingly unstylish. How are we ever supposed to convert people over to sustainable fashion if it's ugly?

Personally, I think we've come a long way in terms of the options for more sustainable fashion. Let me repeat, I said *more* sustainable. Not perfectly sustainable. Aim for perfection if that's your goal, but don't let yourself get overwhelmed. Let's talk about some ways we can transition a wardrobe to be more eco-friendly (many of these overlap with concepts introduced in chapter 3):

Rent. If you go to only a few formal events a year, what's the point in hanging up a fancy dress to sit and collect dust most of the time? Rent formal wear instead! I remember attending a gala in Toronto with a friend who rented her outfit for the night, and it was stunning! She was getting compliments all night long, and the next day she could return it. Most people wouldn't want to wear the same outfit to the next event, anyway. Check out Rent the Runway in the U.S. or Beyond the Runway in Canada, among others!

Borrow. This one is also fantastic, especially if you don't have a rental option nearby. Borrow an outfit from a friend! Clearly it helps to be the same size, so that can be a limiting factor, but think of friends and family members who are a similar size to you and talk to them about borrowing and sharing outfits once in a while. It helps to mix up your wardrobe without costing a dime! Just be sure to wash the clothes before you return them, or your friendship could be short-lived.

Swap. Hold a clothing swap with friends and family. This is perfect for passing along your unwanted clothes, but it's also a great way to get rid of kids' clothes! Local organizations also often hold clothing swaps, so keep an eye out for those to get more variety. Good times to hold clothing swaps tend to be as the seasons change, and times of the year like spring cleaning.

Shop secondhand. Before automatically looking to buy brand-new, check out secondhand shops. From charity stores to consignment shops, and plenty of online secondhand stores, there are a lot of options to buy preloved clothes and accessories. Sometimes you don't even have to pull out your wallet, because you could join a Facebook group where everything is free! This is typically a more affordable option, and it's way more sustainable because the product already exists. I know some people might not be down for buying preowned clothing and accessories, whether because of perceived hygienic concerns, perceptions around social status, or any other reason; there can be mental barriers to thrifting. Keep in mind that you can thoroughly clean everything you buy! Plus, you don't have to start buying undies at Goodwill (I've personally never purchased undergarments secondhand); go for something less intimate instead, like a handbag.

Buy sustainable/ethical. If you can't find what you need for free or secondhand, then find options that are sustainable and ethical, if possible, using the Less but Better Method. Look for brands that have very clear values and are fully transparent about how and where their clothes and accessories are made. For example, I love that the clothing brand icebreaker has an annual transparency report as part of its commitment to build a sustainable and ethical brand, and it's available to download from the company's website. Similarly, Encircled creates comfortable, stylish, and timeless pieces, and its website is transparent about its entire supply chain—and everything is made in Canada! The more information you can get, and the more a company provides answers to your questions, the less it is trying to hide. That transparency tends to be a good signal that it is legit, but you'll have to make that choice for yourself. Also, look for brands that stand by their products, like Patagonia, an outdoor clothing and gear company. Patagonia now has a program called Worn Wear, which helps people get their

Patagonia clothes repaired, sells secondhand Patagonia clothes, and even offers a line of Patagonia ReCrafted clothes, which are clothes made from other clothes to keep them out of the landfill. Not only does Patagonia stand by its stuff, but it is also extending the life of its clothes through repairing, upcycling, thrifting, and recycling. Now, if only we could get all brands on board with this approach!

Choose quality clothes. Adding high-quality clothes and accessories to your wardrobe is always recommended, regardless of whether the clothes are secondhand, free, or brand-new. What does quality mean? Typically, you can tell by looking at what the material is, feeling the weight of the fabric, assessing the construction, and seeing if there's any pilling or tears. Keep in mind that price doesn't always equate to quality. Just because something is expensive doesn't always mean it'll be good quality, so take a closer look before making a purchase. Here are some tips to find higher-quality clothes:

- **The fabric feels thicker.** Typically, thicker fabrics will have more longevity by withstanding more washes. While thin fabric is great for hot days, if it's too thin, it likely won't last long. Not sure if it's too thin? Check the garment's transparency; if you can easily see through it, there's a chance it might not have a long life expectancy.
- **Good stitching is well reinforced.** Have a close look at the seams, and if you see any loose threads, that's a quick sign of poor-quality construction. Gently pull at the seams, and if you see gaps appear, that's another red flag. High-quality garments will have extra rows of stitching, making gaps nonexistent.
- **The construction is more complex.** High-quality clothes are constructed to fit well on people. That might include elements such as an

extra piece of fabric at the shoulders in items like collared shirts or darting (tapering) to give tops a better fit. Other elements of better construction include liners in pieces like dresses, skirts, and jackets, to help them fit better.

- **Patterns match at the seams.** If you're buying a patterned garment, higher-quality pieces will be more likely to match at the seams.
- **There won't be any raw hems.** Unless the garment is specifically designed to have a frayed edge, it should be hemmed properly, with the fabric folded over and stitched well. This will give it longevity; a frayed edge is more likely to unravel.
- **Shoes are stitched, not glued.** When it comes to shoes, stitching tends to beat out glue for longevity and speaks to higher-quality construction.

The ability to find better-quality clothes is a skill that develops over time. You'll learn from experience, particularly if you buy an item and it loses its shape within the first few washes. That's a sure sign to avoid similar clothes, and perhaps that brand completely. When you find clothes that fit your preferences, maintain their shape, and still look great after many wears and washes, then those brands will probably become your tried-and-true go-to brands for future purchases.

Buy less but better. If you find it expensive to buy better stuff, the best option is to buy less in the first place (the Less but Better Method). Even though fast fashion has gotten very cheap, we're spending more on clothes (in total) than ever before! We're just buying more stuff at lower prices and lower quality. According to journalist Elizabeth Cline, "Our closets are larger and more stuffed than ever, as we've traded quality and style for low prices and trend-chasing. In the face of these irresistible deals, our total spending on clothing has actually increased, from $7.82 billion spent on apparel in 1950 to $375 billion today."[4] Before you criticize the price of high-quality clothes, it might be worth considering how much you've spent on fashion in the last year to figure out if you can allocate your funds to fewer but better clothes moving forward.

By following the steps outlined above, you'll have a capsule wardrobe before you know it! What is a capsule wardrobe? A concept pioneered by Susie Faux and then popularized by Donna Karan in the 1980s, a capsule wardrobe is a set of timeless and high-quality closet essentials that can be worn year after year.[5] These pieces, which I would call basics (think your go-to jeans, skirts, blouses, and jackets), can then be spiced up with seasonal pieces and different accessories. Keep this concept in mind as you curate your closet, because with a solid set of closet essentials, you won't have to spend too much money or generate too much waste by adding a few pieces now and again and accessorizing to change it up!

Keeping in mind the idea of longevity, let's chat about how you can make sure your capsule wardrobe lasts season after season. There are two main elements to incorporate: caring for our clothes and repairing them.

Caring for clothes: To make sure our clothes last, we need to care for them. That means following the washing instructions on the label, avoiding harsh detergents, and storing the clothes properly. Also, we don't need to wash our clothes after one wear; it's way more eco-friendly and good for our clothes to wash only when totally necessary. That will vary depending on what your clothes are being worn for; athletic wear that you sweat in will likely be washed more often than your jeans, for example. And new parents, of course, will probably be doing laundry what seems like 24/7 and are exempt from this part of the discussion! Beyond washing, be sure to fold your clothes nicely into drawers, rather than stuffing them in or leaving them lying out on the floor for your pets to snuggle on. A little love goes a long way! If you're hanging clothes, be sure to give them space, instead of cramming too many pieces so close together that you have trouble putting your hangers back in the closet (yes, I've been there). Let them breathe!

Repair: These days, when something is broken, we tend to replace it instead of repairing it. When you start investing more financial resources into your clothes and accessories, though, it starts to make more financial sense to repair rather than replace. Regardless of how much you spent on your clothes (they could have been free, at a swap!), repairing them will give them a better life and be better for the planet. Find a good tailor, or perhaps develop some of your own sewing skills to keep clothes in your closet and out of the trash. Same goes for shoes; look after them, too! When they start to get worn down, you can get them resoled. I've done this plenty of times, and when I do, my shoes come back looking brand-new! While you're at it, get yourself a good shoe maintenance kit, with items like shoe cleaner, polish, conditioner, and accessories like a brush and a cloth.

Become an #OutfitRepeater

I've been an outfit repeater my whole life. As is pretty much everyone I know. Why am I even writing about it, then? Because social media has put pressure on us to show up in new outfits in every post. This is impacting young people the most; 41 percent of all eighteen-to-twenty-five-year-olds feel the pressure to wear a different outfit every time they go out.[6] Yikes!

Time to lay off the social media, right? Whether or not you feel pressure to not wear the same thing twice, here are some tips to keep things fresh in your closet:

Mix and match: You might be wearing all the same things the same way without realizing it, so if you want to freshen things up, add a belt to a button-down shirt or try outfit combinations you've never experimented with before.

Add accessories: Belts, jewelry, scarves, jackets, and so many other additions can freshen up the plainest of T-shirts. Go ahead, get wild!

Swap with friends: Why does your wardrobe have to be limited to your own closet? Mix things up by sharing, borrowing, and swapping with friends! Or by sneaking things out of your older sister's closet, but you didn't get that advice from me. . . .

Aside from keeping your style fresh, it's about a mindset shift, too. It's about being confident in your own skin and loving your own appearance. Outside validation? Sure, it's nice, but your main source of validation should come from within. Love yourself first. Remember, most ads are preying on our insecurities. Be so secure in yourself that you play by your own rules! You're in charge of your life.

People are drawn to confidence and personality, so express your personality with confidence instead of pulling out your wallet for a new outfit every time you want validation. This advice is as much for me as it is for you, because we can all stumble onto the consumer treadmill. With your well-developed sense of style and proud #OutfitRepeater vibe, you'll be glowing with confidence, and everyone will want in on your secret; share it with them! We need more people on board.

Reflection and Checklist

I recall a barbecue I had with a small group of close friends a little while back. The summer had been super hot, but that day was an exception, and the evening was cooling down to the perfect temperature. My friend Jenna popped on a cute floral cardigan, and I told her it looked really nice on her. She thanked me and told me she had bought it for her recent vacation to Jamaica. Jenna then proceeded to explain that it was fast fashion, as if to somehow preempt the conversation from heading down the path of sustainable fashion. All of our friends know that I promote sustainable living, and when I first made the transition, people treated me a bit like the plastic police. I'm not here to monitor anyone, and I'm certainly not here to judge people's fashion choices.

I'll be the first to admit, there are limitations to sustainable and ethical fashion, in terms of both style and price. It's really a balancing act between the two. My recommendation to you, as I mentioned before, is to buy your clothes with longevity in mind, fast fashion included (even if that is unpopular advice coming from a sustainability advocate). Much of fast fashion is bought one week, then tossed the next. The thing I know about Jenna, though, is that she keeps her clothes for a really long time. She looks after them and even still has some pieces that she wore in high school! There's no need to feel bad if you know you're going to love and care for your clothes. Ultimately, it's about doing better when and where we can.

I'm no saint in the fashion department. It's not like I've never bought from companies that are notorious for pumping out fast fashion. A few of my most treasured pieces over the years have come from these gigantic fashion retailers—this includes a wool jacket I bought ten years ago and still wear today. But even back when I would buy fast-fashion brands, it was never with the intention of wearing an item once before tossing it. If you plan to buy fast fashion, buy it for the long term, and choose pieces that seem like they'll last way longer than one or two washes.

Fashion is so personal. It can reflect our mood, personality, confidence, finances (or projected appearance of financial status), culture, religion, and so much more. It can be fun, as we can make our clothes an outward expression of our beliefs, values, and character. It can also feel limiting, especially if our budget can't stretch as far as our desire to build a dream closet. The best thing we can do is make what we have work for us at this point in time and, as with everything else, create wish lists and write down goals to work toward. Remember, it'll always be a work in progress. Your closet doesn't have to be perfect today. It's that desire for perfection that sends people to fast fashion. And speaking from experience, this doesn't fill the true void.

This month might feel deep in terms of the personal assessment and decluttering that go with it. Let's turn it around and think of it as empowering! I love making well-thought-out and informed choices about my wardrobe. That might not be realistic for every purchase, but certainly it's possible to be more mindful most of the time. Find the balance that works for you, and no matter your personal circumstances, you can add your personal touch to your style. Here's a quick summary of this month's actions:

- Understand your reasons for buying fashion, and learn how much of it you're buying and how much you're spending on it
- Declutter and clean out your closet, keeping only what you know you love and will wear
- Develop your personal sense of style, and avoid fleeting trends
- Create a wish list for your closet, and stick to it for future purchases
- Shop less, buy better, and swap/share/borrow when possible
- Care for your clothes, and repair them when necessary
- Develop confidence in yourself and in your style, so that it doesn't need to be propped up by frequent fast-fashion purchases and external validation

Chapter 7

The Subtle Art of Refusal

Discover the joy of saying no to stuff

Thanks, but No Thanks

For most of my life, I was a "yes" woman when it came to all of the things. Do I want to bring home those mini shampoo and conditioner bottles from the hotel? *Yes!* Do I want that free bag of flyers, branded pens, and extra notepads from the conference? *Yes!* Do I want to bring home the wedding favors? *Yes!* Do I want to spend an extra twenty-five dollars on makeup in order to get the "free" beauty kit full of sample-sized products? *Yes!* Do I need a bag with that? *Yes!* Do I want a receipt? *Yes!* Do I want to subscribe? *Yes!* Do I want that rewards card? *Yes!* Do I want that unattractive figurine gifted to me by a co-worker? *Yes!* Random stuff was flowing in left, right, and center—because I was letting it in! I became inundated with crap. Are you?

It's hard to say exactly why I had no problem letting all of this stuff into my life. I think it must have been a combination of social norms (saying no can come across as rude) and the thrill of getting "free" stuff. These days, however, I very closely scrutinize what I say yes to, because I've come to recognize the value of my time and energy—oh, and storage space. Plus, it's wasteful to collect all of these things, only for them to end up in the trash. As we've explored, it takes energy and resources to create all this crap in the first place, and we create demand for it by welcoming it into our lives. Eventually, though, its home will undoubtedly be in a landfill.

Saying yes to all the extra miscellaneous stuff requires energy. It can be exhausting! Think about it this way. When I brought home any of that extra crap, from the miniature hotel toiletries to the "free" conference goody bag, those things had to get sorted and put away. And after that they were taking up real estate in my home, collecting dust. If we're honest with ourselves, we know that most of these things are items that we will eventually declutter, once we recognize them for what they truly are: unwanted crap. Turning off the crap tap—in other words, refusing this stuff in the first place—will save you the time and energy of bringing it home, putting it away, caring for it, dusting it, moving it around when you need something else, and eventually decluttering it. This is precious energy! Wouldn't you want to spend those hours of your life doing other things?

That's exactly what I discovered: I didn't want to spend time and energy collecting, sorting, and throwing away unnecessary stuff that I could have refused in the first place. Let me repeat: *in the first place*. What's your best shield against these things? Saying "No, thank you," when freebies are coming your way. Don't be a free-flowing "yes" person like I was; instead, more closely examine whether you want these things to enter your life and your home. What's another ingenious tactic? Avoiding scenarios in which you know you'll be receiving something that you don't want or need (like *another freakin' vase* from Aunt Betty) by being up front with family and friends. And while I don't have children, I've heard that there's a lot of crap that comes home with them, too!

You might be thinking that it'll be tough to go from a "yes" person to a "no, thank you," person, and I get that. It's a change, and change can be hard sometimes, both for us and for our loved ones. This is especially true when people have always known you to behave a certain way, but they'll get used to it! Just be open and honest about your preferences with everyone. Once you provide an explanation, most people will get it. Sure, Aunt Betty might be a bit upset, but perhaps that means you need to have a deeper, more meaningful conversation with her about your newfound ways.

This chapter is all about how to master the subtle art of refusal without losing your friends and upsetting your family in the process. It's all about setting new boundaries and moving beyond the awkward-AF conversations. Here are some simple steps to becoming a pro in the subtle art of refusal: say "no, thanks" to random stuff; develop new boundaries with family and friends; manage gifting and reduce waste over the holidays. When you cut the crap, you get your life back. And the landfills will thank you, too!

Say "No, Thanks" to Random Stuff

Junk mail showed up daily in my mailbox for years. It didn't occur to me for the longest time that I could refuse it. This is a fine example of a crap tap, where we receive stuff that we don't need or that could be available to us in a less trashy format. Here's how to turn off the junk-mail crap tap and some others:

> **Junk mail:** If your flyers go straight from your mailbox to your recycling bin, then it's probably time to refuse junk mail. The approach you need to take to decline junk mail will vary depending on where you live, so be sure to ask your postal service how to avoid receiving flyers and various other advertisements in your mailbox. That information is

probably available online. Where I live, all I need is a "No Junk Mail" sign above my mailbox.

Bills and paperwork: These days, most paper communications, like bills and other notifications, can be sent to you digitally. Avoid paper and set up a new online system by opting for digital bills.

Business cards: Try finding ways to connect online, such as using LinkedIn, instead of taking business cards. When someone hands me a business card, I say, "Can I add you on LinkedIn instead?" It surprises most people at first, but once the idea registers, I've gotten comments like "That's such a good idea!" and "How smart!" as if it were revolutionary! I also take photos of business cards with my phone, which is quick and easy.

Freebies: Freebies come in all forms. Choose to refuse hotel toiletries, bonus gifts with purchases, free samples, goody bags, branded promotional products, etc. Don't be afraid to say "no, thank you," directly in situations where people are trying to hand off free stuff to you. It might feel awkward, but in most situations, you don't know them personally, so you don't have to worry about offending them. Plus, when you're polite about it, it's their problem if they don't take the refusal well. If the conversation lasts a little longer, you could explain that the refusal is part of your effort to live a low-waste lifestyle.

Physical subscriptions: A lot of newspapers and magazines can be read online, so it might be worth making the switch if you're still receiving the physical versions of various subscriptions.

Kids' stuff: Kids come home with a *lot* of stuff, especially once they start going to school. From artwork and homework to forms from teachers, it can be overwhelming. Based on what I've been told, parents must be stealthy in their approach. For artwork, refusing isn't exactly the right tactic, so let it come home. Then, together with your kids, proudly display their favorite pieces, and discreetly discard the rest. As for paperwork, it's worth connecting with your kids' school to find out if you can transition to a digital format. For everything else, make it a priority to teach children the value of less as they grow up. Everything you're doing and transitioning toward with sustainable living they'll pick up from you. Teaching them how to say "no, thanks," when appropriate, will help reduce the amount of stuff that comes home with them.

It may take some time to get used to this new approach of refusing, but you'll get the hang of it quickly. It's an empowering experience, because this is a subtle act of rebellion, and suddenly you might feel like you're a part of a secret club that not many people know about. At least, perhaps, that might be how I feel sometimes and I'm now projecting that through this book. Welcome to the club! But it shouldn't be top secret. Instead, spread the word!

Much of refusing is really about planning ahead and setting up systems that prevent the inflow. But a small amount of planning and setup goes a long way. Now that I've prevented junk mail from showing up in my mailbox every week, I no longer have to collect it and transfer it directly into my recycling bin.

"It's Not You, It's Me"—Develop New Boundaries with Family and Friends

If we're honest, none of us actually wants to be in a situation where we say no to someone directly. The subtle art of refusal is really about having these important—although sometimes awkward—conversations in advance so you don't have to say no at all. As I reflect on my experience working toward minimalism and living less trashy, I recognize that this was a life *transition*.

It didn't happen immediately, nor should you expect your family and friends to understand the new version of you after a single conversation. Before you hop into a quick chat with your loved ones about your new and improved preferences, it's worthwhile to simmer your own expectations of what the outcome may look like. Chances are, you'll need to have some repeat conversations for the point to get across. Sometimes the point you're trying to make may never be fully digested or understood on the other end.

Let's begin. There are two essential ingredients that make up the subtle art of refusal. The first is having conversations with your family, friends, and perhaps even co-workers around your lifestyle changes and new personal preferences. That may be followed by more direct and up-front conversations in which saying no is necessary. We'll cover optimal strategies for those awkward conversations below. For now, here is the two-step process for these conversations.

1. **Discuss your new values.** Consider the old version of yourself and ask: *What is different about me now?* People who know you will have certain expectations about your lifestyle habits based on what they've seen you do up to this point. If you're transitioning to a less trashy lifestyle, it's because your values have changed. Perhaps you previously valued convenience (which resulted in

the disposal of many single-use plastics, like to-go coffee cups) and now you value sustainability and keeping plastic out of our oceans. Your job in this conversation is to fill in the gap between what people expect of you (your old behaviors) and what they see you doing now (your new lifestyle choices) by explaining what your new values are and why you have them.

2. **Explain how your new values inform your preferences.** Now you can benefit from taking it a step further. Having already discussed your new *values,* you've laid the foundation for future conversations around your new *preferences.* Going back to that coffee cup example, let's say one of your co-workers, Gary, brings you coffee from the local coffee shop in a disposable cup each morning. It's become a ritual, because you always bring Gary baked goods. However, because of your new values, you'd prefer coffee in a reusable travel mug. Since you already had a great conversation with Gary the other day about your new values, it won't come as a shock to him when you eventually (and kindly) explain that, although wholly appreciated, it's no longer a great fit to get coffee each morning in a disposable coffee cup. The important thing to do when you identify a new preference is to follow that up with an alternative solution that everyone will be happy with. In this case, you could suggest that you both make coffee at work in your own reusable cups. Whether or not Gary takes you up on your offer doesn't matter, because you've shared your new preference, based on your new values, and offered a new solution. That puts the decision making back in the other person's court, and they get to enjoy some ownership on the path forward. It's a win-win.

This two-step conversation will set you up for long-term success. Perhaps it'll take some time for others to transition with you (Gary might forget and, out of habit, still grab you a coffee), and that's okay. Kind and respectful repetition may be required. That said, some people may never change their ways, no matter how many times you outline your values and preferences. I believe that's okay, too. Grandparents, for example, can get a free pass in that department! If your grandparents, or anyone else, for that matter, simply can't or won't change their ways, then focus your efforts where they will be more fruitful. Aside from these outliers, most people in your social circle will respect your wishes, and you'll eventually find that people get used to the new you!

Now let's broach those more awkward, up-front, in-the-moment conversations. While saying "heck no" to something you don't want is fine for your internal dialogue, what comes out of your mouth should probably be a bit more tactful. Chances are, you'll end up in circumstances in which you need to think on the spot about how you'll say "no, thank you." Here are some polite ways to say no to unwanted things:

- "Thank you kindly for your offer, but I know I won't use this, and I don't want it to go to waste by accepting it."
- "Thanks for the kind gesture, but I don't have any space in my home for this item. Is there someone else who might benefit from it instead?"
- "Thanks so much for thinking of me. I already have one of those. That said, I know Barb has been looking for one for ages! Have you considered offering this to her instead?"
- "I'm working toward a more minimalist lifestyle, so I will have to decline, but thanks for thinking of me."

- "Thanks for thinking of me! I don't have a need for these things. Have you considered donating these unwanted items to a nonprofit organization?"
- "Wow, thank you. That's very generous. I already have too much stuff, so I have to pass."
- "Thanks for your generosity. I already have one of these. Do you think this could be passed along to someone who needs this more than I do?"

It can be hard to say no, but it's better (like, more emotionally freeing) than saying yes when you really mean no. Remind yourself of the true cost of saying yes (your time, effort, energy, etc.), and this may help reinforce why it's worth having a quick awkward moment that will soon pass. Chances are that the more you do it, the more comfortable and confident you'll be in your decision to say no in the moment. There may be moments, however, when it would be far too uncomfortable to say "no, thank you," and that's okay, too.

How to Handle Gifts

Growing up, I loved receiving (and giving) gifts. At Christmastime, the more gifts under the tree, the better! While I didn't live in an overly religious household, we still celebrated Christmas, among other holidays. Our version of these holidays was very commercialized, rather than religious. The focus for our household was on family time and gift giving rather than praying and going to church.

Putting up our Christmas tree each year brought me so much joy! I still remember the excitement of pulling out our tiny plastic tree; it was only about three or four feet tall, so we'd put it on our wicker side table to give it some extra height (and make room for more gifts underneath, of course!). We'd add a string of lights, as well as a mix of lovingly homemade and store-bought ornaments that we had collected over the years. My favorite thing to do was to sneak out in the middle of the night, once everyone had gone to bed, to sit and watch in awe as the tree glowed in contrast to the nighttime darkness. To this day I'm still surprised I never caught Santa in action!

This special quiet moment wasn't solely focused on what Santa would bring me. It gave me time to reflect on my happiness and joy, even though our family life was far from perfect. Thinking about it now, I realize those quiet moments were really about gratitude. I was grateful for the moments with my family, for the opportunities that I enjoyed in my life, and for the gifts, of course, too. There's magic in being young and naive, and it was magic that I felt while gazing at our beautiful Christmas tree in those peaceful, twinkling moments. Our tree was modest, but nonetheless my heart was bursting with delight.

Gift giving is part of all cultures and has been an act of reciprocity between humans for thousands of years, an integral part of celebrations, religious ceremonies, and more somber functions like funerals. But like everything else, gifting has become highly commercialized; not surprisingly, marketing has upped the ante when it comes to what's expected on these special occasions. The pressure this puts on consumers has contributed to high levels of debt and creates the stress of trying to impress others and show them our love in gifts—some of which we can't afford, which, of course, creates more stress. Americans took on an average of $1,381 in holiday debt in 2020.[1] Why is it that we feel the need to show our love by amassing debt? Something has got to give!

Over the years, I have, whenever possible, become particular about the gifts I receive. I'm well aware that sounds a bit ungrateful, but I feel it's better this way in the long run. Oftentimes gifts are chosen by the gift giver, perhaps a friend, family member, or even co-worker. Maybe they knew what you wanted, maybe they didn't, or perhaps they just found something they thought you'd love.

The problem is, we often *don't love* the gifts we receive. We express our thanks, only to bring home the gift and hide it away in the deep, dark depths of a cupboard. Instead of receiving gifts that you don't want, why not give people a bit more guidance? My suggestion for making sure you receive gifts that you will actually use is to have a wish list (like a wedding registry) and to share it with loved ones or let them know in advance that you don't want any gifts at all. You can make such a list for occasions like birthdays, baby showers, and bridal showers.

While some people feel that creativity is removed from the gift-giving process with a defined wish list, there are still ways to make it more fun, like leaving some part of the gift open-ended, such as the color, the style, or the type of experience or consumable. Similarly, it's worth asking friends and loved ones what they might love to receive as a gift.

Not all occasions are necessarily appropriate for sharing a wish list for gifts. Some occasions where it might be inappropriate are housewarming parties or adult birthday parties. In those situations, you could say something like "No gifts necessary, but if you would still like to bring something, please make it a dessert or bottle of wine we can enjoy together." The expectations are clear, and you'll avoid unnecessary and unwanted gifts with this approach, plus, consumables don't create clutter in your home!

If your efforts are not fruitful (or, ahem, are too fruitful), you may need to reiterate your preferences over time (as we addressed above with regard to setting boundaries). With certain people, like Aunt Betty, you might have to give up altogether, knowing that they aren't likely to change their ways. If that's the case, here are some techniques to handle unwanted gifts:

Return it: If you've received the receipt, or even if you haven't, you can try to return it for a refund or gift card from where it was originally purchased.

Regift it: Give it as a gift to someone else, but be sure you know they'll want it and will use it. To avoid awkward conversations, try to regift it in a different social circle from the one where you received the gift. For example, if you received the gift from a family member, then it might be best to regift the item to a friend, and vice versa.

Sell it: If it's worth the time and energy, try selling it.

Give it away: You can either give it away for free using online websites like Craigslist or Facebook groups or donate it to a charity that would use or resell it.

The act of receiving a gift is normally the most important part of the exchange, which is why refusing it in the moment doesn't work in most situations (this is what could make you seem ungrateful). That said, once you receive it, the gift is now yours to do with as you please. There's no point in letting an item collect dust in your house; it might as well find a new home where it will be used and loved. Thank the person gifting the item, and later send it off to a new home, guilt free.

And as for the other half of the equation—giving gifts to your loved ones that will put a smile on their face without taxing our planet—read on.

Navigating Holidays

In the United States, people throw away about 25 percent more waste between Thanksgiving and Christmas than the rest of the year; and if each American family wrapped three gifts in reusable materials, it would save enough paper to cover 45,000 football fields.[2] Between food waste, gift wrap and bags, tissue paper, cards, ribbon, boxes, and more, the holiday season is *trashy*. Refuse to be part of those statistics and find fun ways to reduce your waste instead!

When I was growing up, it occurred to me how strange it was to spend time wrapping gifts, only to unwrap them on Christmas Day and throw out all of the nice paper, ribbon, and bows. I questioned it and saved a bow or two for the next year but at the time didn't manage to consider a new way of doing things that would avoid this practice altogether. That has changed, of course, and Hubby and I have focused our efforts on reusing more and tossing less over the years. To help with our newfound love of reusing, we purchased a vast quantity of reusable cloth gift bags to wrap gifts during the holidays. It makes "wrapping" quick and easy, and once the gifts are unwrapped, we put the gift bags away to be reused the following year.

While I consider myself minimalist-ish, I don't mind holding on to items like these reusable bags, because I'd rather store them than create waste. Our family members know that these bags are to be kept and reused for the following year, so there isn't any expectation that they will keep them and take them home with their gifts. For friends, we give the reusable bag as part of their present, and they can reuse it for their gift giving, too! Here are some ideas to avoid waste during gift-giving occasions:

Reusable bags: You can either make or buy reusable gift bags. They can be made of cloth or paper, but the more durable they are, the longer they will last.

Newspaper: Newspaper, magazines, and other materials that will just get tossed anyway could be repurposed as wrapping paper.

Save bows and ribbons: Instead of throwing bows and ribbons in the garbage, save them to be reused for the next gift.

Or you can avoid giving physical gifts altogether! Let's consider some gifts that aren't actually things at all:

Gift experiences: Who doesn't love making memories? Give gift experiences that they will love, whether it's as simple as a pair of movie tickets or more thrilling, like a day on the ski slopes. Here are some ideas:

- Date night
- Digital subscriptions (like Netflix or Spotify)
- Gym memberships (digital or in-person)
- Memberships to art galleries or museums
- One-on-one lessons for specialty hobbies, like knitting, cooking, or chess (in person or online using Zoom)
- Spa day and spa services like a massage or manicure

Gift favors: From babysitting to shoveling snow, there are plenty of favors to go around. Put your skills or time into action with gifts that people will love! Here are more examples:

- Cleaning
- Cooking/baking
- Construction/renovation assistance
- Massages
- Shopping trips (like grocery runs) or other errands
- Yard work

Gift consumables: Like experiences, consumables are great because they won't hang around in someone's house indefinitely (like, flowers can eventually go in the compost bin!). Anything from a bottle of wine and baked goods to a home-cooked meal—there are endless options!

Gift donations: Gifting donations is a great way to give back to your community! You can either ask the person you plan to gift to what organization/charity they would like a donation to or choose one based on your knowledge of their favorite causes. People are often excited to receive a gift that has a greater impact.

Reflection and Checklist

It can really feel like we're going against the grain when we refuse freebies and decline gifts. There's an art to doing it subtly, in a way that doesn't offend people. Mostly that involves kindly chatting with friends, family, and co-workers in advance about what your personal preferences are around receiving gifts. Some people will get it, and some won't. Most of the time people will be open to your request and respect your wishes.

I personally still love giving and receiving gifts, but my approaches to giving and preferences around receiving have changed, and I've done my best to communicate that to my family and friends. It's all about setting boundaries—and once you see how rewarding it can be to declare your wishes and have people respect them, you might find yourself setting all kinds of new, healthy boundaries in your life. When you're ready to implement the subtle art of refusal, here's a recap list of actions:

- Turn off the crap tap by setting up systems that avoid things like junk mail, paper bills, and unwanted gifts
- Decline unnecessary freebies like hotel toiletries and gift bags
- Set up new expectations with family and friends around your personal preferences for receiving gifts
- Put your money where your mouth is, and give gifts that reflect your values (like experiences and consumables)
- Let go of unwanted gifts without guilt by finding new and better homes for them

Chapter 8

Family and Friends

*How to not be a know-it-all and
tell-everyone-how-to-live-their-lives kind of person*

Sustainability Rock Star

My newly repurposed jars were freshly washed and dried and tucked away in my backpack for their first trip to the bulk food store. I laced up my sneakers and walked out the door. Not only was I going grocery shopping with my own containers for the first time, but Hubby and I were walking there. Watch out, world, I was a sustainability rock star! The bulk store I went to had a reusable container program, which allowed shoppers to use their own containers and jars for anything from chocolate to rice to all-natural, organic, fancy-shmancy peanut butter. I just had to get my containers weighed and examined by staff first; then I set off to fill my jars! It felt so good, until someone on my social media account told me that my low-waste shopping wouldn't solve the world's problems. Womp, womp. Someone had to just rain on my parade, and there's a good chance you'll experience that, too.

Meanwhile, my husband and I had been avid Costco shoppers for years (we still are, we just shop differently now), which, sadly, resulted in tons of excess disposable packaging. Even though we had been keen on sustainability for a while, we never really questioned the items we tossed into our trash bin or recycling box each week. This was a whole new world to us, and now we had the eagerness of new converts. Since you're reading this book, there's a good chance you're feeling the same way. The thing is, though, just because it's important to *you* doesn't mean it will be important to your friends, family, or co-workers. They haven't had the same awakening as you, and that's a rocky road that can be tricky to navigate at times.

Hubby learned alongside me, although maybe not to the same extent, nor with the same enthusiasm. It was still helpful to have him aware of the new information I was gathering over time, because then he would fully understand why changes were happening in our home. When he understood *why* we were making a change, I found that the change went down easier.

Take toiletries as an example. Most people find the hair and body products they like and stick to them indefinitely. When we started reducing waste in the bathroom, as discussed in chapter 5, it was like starting from scratch, experimenting with new products. If Hubby hadn't understood the purpose of these changes, he might have been less inclined to try the new shampoos, bar soaps, and other concoctions, lotions, and potions I brought home.

Eventually, we landed on a plastic-free shampoo bar that we both use and love, a simple and sudsy soap bar instead of bottled body wash, and metal safety razors with recyclable metal blades instead of plastic razors with disposable blades. I didn't force any of these swaps on him, but having sat beside me as I sputtered through documentaries, he was happy to adapt to and adopt my newfound ways. It was a slow and steady transition, which probably also helped.

I doubt it's always this easy in other households. There's a good chance you will face resistance from loved ones and friends about making these less-trashy-lifestyle transitions at home and on the go. While it would be wonderful if everyone could just see the world as we see it and adopt our new practices, people don't operate that way. The motivation for lasting change needs to come from an internal fire. That doesn't mean we won't be able to influence people around us—we can and should, in an ethical way, of course! Whether or not our loved ones join in, family support goes a long way when it comes to making such important lifestyle changes, because it can reaffirm our choices.

That's what this chapter is all about: how to get friends and family on board while not being that know-it-all telling everyone how to live their lives. Let's break down how to walk that line: you do you, first; invite others to learn with you; lead by example; and learn what to do when your loved ones aren't on board. This approach will, hopefully, inspire and influence the people around you for the better! If not, move on.

First, You Do You

Did you know that seabirds like eating ocean plastic because it smells good to them?[1] Apparently, scientists have discovered, plastic, once covered in natural growth like algae, suddenly becomes super appetizing to seabirds, and that's why they eat it. Interesting, right? I learned this because I take the time to research, read, watch documentaries, and focus on growing my knowledge. You're not going to magically pull *real* information and data out of your ass, and I can't either!

The point is, know your stuff and educate yourself! It's not a one-and-done activity; it's a lifelong learning kind of thing. New information, data, and research are becoming available all the time. It's necessary to keep up to date because life changes, and so does our impact on the planet. Stay aware of what's going on. So before you go trying to school anyone (which you shouldn't do anyway), school yourself. You do you, first. Do the research. Do the work. If anyone wants to challenge you on your beliefs, lifestyle changes, or choices, you can back up your position with hard facts and evidence.

Keep in mind that you don't have to be an expert, like an environmental scientist, biologist, or chemist—the people on the ground who are collecting all the data and sharing the important statistics and information with us normal folk. If you fall into the category of expert, that's great! You already have a wealth of knowledge. The rest of us, however, have a little extra work to do to make sure we have some kind of idea about what's going on in the world.

You don't have to know everything, but even when you know just a few things, people will be impressed. You can start a conversation with "Did you know that only nine percent of plastic ever produced has been recycled?"[2] And *bam,* you know what you're talking about, and you can inspire others. Here are a few extras you can have up your sleeve if anyone corners you demanding details:

- In 2019, the world generated 53.6 million metric tons of electronic waste, and only 17.4 percent of this was officially documented as properly collected and recycled.[3]
- About one-third of the food produced in the world for human consumption every year—approximately 1.3 billion tons—gets lost or wasted.[4]
- Per capita food waste by consumers is between 95 and 115 kilograms per year in Europe and North America, while consumers in sub-Saharan Africa and South and Southeast Asia each throw away only 6–11 kilograms of food waste per year.[5]
- Humans consume more than one million single-use plastic bottles per minute.[6]

- There is no limit to how many times aluminum can be recycled.[7]
- Recycling plastic downgrades its quality.[8]

Invite Others to Learn with You

Learning with others is a great way to encourage them to modify their life-styles to be a bit less trashy. It's a discreet way to help others see the world you see, without telling them how they should see it. Watching a documentary together that shows how single-use plastics that are dumped into our waterways are killing wildlife is likely to be more effective at convincing people to skip the straw than just telling them they should stop using plastic straws.

Back when I was working a nine-to-five office job, I decided to create a green team because I wanted to help inspire positive change at work. I'll share more detail on that in the next chapter. As part of this green team, we organized events like local litter cleanups and watched the occasional documentary during lunch hour (ordering pizza can help draw a crowd). I specifically noticed an uptick in motivation to reduce our use of disposables (like plates, cups, cutlery, and napkins) during staff potlucks after watching the documentary *A Plastic Ocean*. It always seemed silly to me that staff used disposables when we had easy access to a kitchen with a sink and dish-drying rack. The main issue was getting volunteers to do dish washing, which suddenly was no longer a problem after watching this documentary. It was a bit like magic!

Keep in mind, though, that old habits die hard. People tend to forget about their motivation from a single moment in time like watching a documentary. Eventually most of us slip back into our trashier habits (I noticed this happen at the office, and it happens to me, too). The best way to combat that barrier to long-term improvement is with continuous exposure to new inspiration; be sure to keep it mostly light and interesting, because it can get a bit too disheartening (for everyone, including you) to be constantly bombarded with images of dead whales with stomachs full of plastic. That said, there are so many amazing documentaries to watch, and they always reinspire me to action. While these recommendations will eventually date this book, here is a list of documentaries I enjoy that you might, too! New ones will emerge, so my tip is to keep learning together.

- *A Plastic Ocean*
- *Minimalism: A Documentary About the Important Things*
- *The Biggest Little Farm*
- *Kiss the Ground*
- *This Changes Everything*
- *Stink!*
- *RiverBlue*
- *Before the Flood*
- *Anthropocene: The Human Epoch*
- *Chasing Coral*
- *The True Cost*
- *A Life on Our Planet*
- *Revolution*
- *Food, Inc.*
- *War on Waste*

Lead by Example

Since transitioning to a less wasteful lifestyle, I've eaten way healthier than at any other time in my life, become more mindful with money, and made better purchases resulting in far less buyer's remorse. In addition, I've expanded my community network, made new lifelong friends, and had the chance to chat with people from all walks of life about the global trash issue. This life-style transition has been a pretty cool experience, and it constantly surprises me with new and exciting opportunities, like writing this book.

Here are some of the beneficial "side effects" you may experience as you start to live less trashy:

- Less stress
- A healthier diet
- A healthier weight
- New friends
- More financial savings
- New skills
- Broader knowledge that enables you to teach others along the way
- Positive feelings of contribution

Whatever gains you're experiencing, be sure to share them coolly and casually in conversation with friends and family. Your personal changes may lure others to jump on the bandwagon, whether it's now or down the road.

What to Do When Your Loved Ones Aren't on Board

Try as you might, not everyone will make the switches you think are impor-tant for a better, greener future for us, the planet, and future generations. It's a total bummer; I get that. It's not our job to tell anyone else how to live their lives, and we always have to remember that. No one needs us to become the plastic police. That said, you don't have to give up on it right away. You could, if you feel it's appropriate, have a productive and positive conversation.

I recall when I participated in a panel discussion with a really great group of sustainable-lifestyle educators. One of the most common questions I get when speaking at events is how to get friends and family on board, especially if they don't seem interested whatsoever. A friend of mine from the green community, Meera Jain, had the best response and recommendation I've heard so far. Meera suggested to simply ask what makes them uncomfortable about a specific topic or lifestyle change. Here are some examples:

- What makes you uncomfortable about recycling? or composting?
- What makes you uncomfortable about bringing your own coffee cup? or your own reusable water bottle? or reusable bag?
- What makes you uncomfortable about low-waste living?
- What makes you uncomfortable about shopping secondhand?

They may or may not share their concerns with you. You should be prepared for the plain and simple response that they just aren't very concerned about these environmental issues. People are different; they likely have different views, opinions, and lifestyle interests. While it may be easy to think that we should all care about sustainability because we share this planet and ultimately have the same basic needs, we still might not land on the same page about these matters.

Should your family and friends take the time to answer your questions, then have an open and friendly conversation about it. Don't jump to judge their perspective. This is exactly the moment when you *should not* suddenly turn into Captain Planet, becoming a hero for the environment and throwing all excuses for inaction into the trash. Rather, take an understanding point of view and listen, as it could offer you insights into what they may be perceiving as barriers to the lifestyle. Then, if they *are* interested, you can offer suggestions to help them overcome obstacles.

One issue you might run into is that many people believe individual actions don't matter and won't make a difference. I suspect you believe individual actions matter; otherwise you probably wouldn't be reading this book. I personally believe that individual actions matter, but not in isolation from other important actions taken by governments and large corporations. Nothing on this planet exists in isolation. Yet here are some arguments you might run into:

- "Placing the onus of sustainability on individuals is propaganda so that big, polluting corporations can keep doing what they're doing to destroy the planet."
- "Individuals can't change climate change."
- "Focusing on individual action distracts from government accountability."
- "Why bother? We're screwed anyway."

These statements, at least to me, seem like excuses to justify personal inaction. It also reminds me of the bystander effect, which is a theory that individuals are less likely to help someone in need when there are other people around. Similarly, choosing inaction because it's someone else's (government's, corporations', etc.) responsibility is a form of avoidance. Don't get me wrong, I think it's inappropriate for plastic-bottle-producing companies to put the onus on individuals to clean up their communities through litter-fighting campaigns. But if you recall the fact I highlighted above, that people are buying more than one million single-use plastic bottles per minute, a simple response could be "Wouldn't you agree that consumers are also part of the problem, based on that level of consumption?" This debate isn't new, and I don't expect to resolve the argument with this book. Rather, if you find yourself having one of these conversations, here are some thoughts to consider adding to your discussions.

An individual has influence over others around them. Take littering, for example. If you're around a group of people who don't litter, and you choose to litter, your behavior (an individual action) will likely be frowned upon. In that situation, you wouldn't be likely to litter again. That's because individuals don't live in isolation; humans are part of society, and when our actions spread (think about how Greta Thunberg went from a solo climate strike to creating the largest collective climate strike ever seen), they are compounded, and that's when we see results. Simply put, individuals are part of a collective. That is the human experience. Therefore, individual actions *do* matter. Hopefully that perspective will help if you run into this type of discussion!

Whatever you do, don't push the conversation. Let it happen naturally, and if they don't want to talk about it, then leave it alone. There's only so much you can do to change someone else's mindset; ultimately, it's up to them to do that. If they come around, they know they can come to you to learn more.

Reflection and Checklist

If you try all of the actions outlined in this chapter and your friends, family, and co-workers still don't get it, or perhaps still don't care, that's okay. You're not likely going to convince *all the people* of your convictions. You'll win some, you'll lose some. If you manage to convert *anyone* to more sustainable ways, that's amazing. Celebrate those wins! Remember, just do your thing. You do you, first. Don't let it drag you down if not everyone joins in.

It can be valuable to have the support of your family, which is why it's worth having a conversation about low-waste living with them. If they're not supportive, or are supportive only to a point, there are other ways to find passionate people who will motivate you to stay on track! There is a great community of people out there (online and IRL) who are keen on sustainable living. We'll address how to build your own network in the next chapter.

Whatever you do, don't forget why you got into this lifestyle in the first place. Living with less trash to become more sustainable is commendable and is an incredibly important means of creating societal change. The actions outlined above have the potential to help win over the people in your life. Whether or not that happens, keep calm and keep going. For now, here's a reflection checklist from this chapter:

- Learn about the environmental issues that are important to you and share that information with friends and family when appropriate
- Don't preach; instead, learn together with family and friends
- Lead by example and let others see the lifestyle benefits you're experiencing
- When faced with disinterest, consider asking what makes your friends and loved ones uncomfortable with certain lifestyle changes, but be open and nonjudgmental
- Don't become the plastic police, but do become available for productive conversations
- You do you, first; live your best sustainable life, and others may or may not join
- Find your community (more on this in the next chapter!)

Chapter 9

You're Not Alone

Find community online and IRL

Good for You

When I first started telling people about my decision to transition to a more sustainable lifestyle, I received a chorus of encouraging responses: "What you're doing is great. It's really important." "Good for you!" Despite general positivity, I was a little dismayed that others didn't immediately feel inspired to make changes in their own lives, too. However, I did notice that the more I shared about my lifestyle, the more people around me would slowly, bit by bit, get on board with making a small change here and there.

Despite knowing that it's best to lead by example, I found it wasn't enough to have people cheering me on from the sidelines; I wanted to hang out with like-minded people who weren't merely saying, "What *you're* doing is great." My family has always been super supportive (thank you, if you're reading this), and getting family on board to some degree can help make your low-waste journey that much easier. But I was still interested in finding friends where the conversation would be more like "What *we're* doing is great. *Let's* do more!" Finding and building a community of people who were interested in reducing their waste and taking care of the environment was helpful for maintaining my new lifestyle. I actually enjoyed changing my habits when I was doing it with others. I learned tips and tricks from people further along their low-waste-living journeys than I was. And somewhat surprisingly, it wasn't very hard to find all sorts of welcoming and engaged communities on the internet, particularly via social media platforms like Facebook Groups and Instagram.

The game changer for me was documenting my lifestyle evolution on a public social media account. That way, people near and far could come along for the journey and share their experiences, too. My social media account led to meeting new real-life friends, and before long I was speaking at all kinds of events. I initiated a green team at my then office job (organizing litter cleanups and documentary lunch-and-learns) and I became a public liaison committee member for waste management in my city. Over a two-plus-year period, I made a whole bunch of new friends, grew my professional network within the green-living community, and became known as a specialist in sustainable living through my blog, *The Zero Waste Collective.*

Developing these communities helped me feel that I fit in somewhere. I felt a lot less like I was swimming against the current of what everyone else was doing, because I found a community that was moving in the same direction I was. And you can, too! You don't have to go so far as to start doing speaking events or writing a book to find like-minded folks in your community and online.

Are you ready to find your community? Whether you're an introvert, an extrovert, or anywhere in between, you can find people and join conversations about sustainable living all over the world! Do it from the comfort of your living room, in your pajamas, or go out and attend events, marches, and conferences. Whatever level of engagement you want, it's out there waiting for you to grow a kick-ass community of like-minded people, build your online network, and make friends in real life. Once you start doing some or all of the options below, meeting people and making friends will be inevitable!

Build Your Online Network

The very first photo on my very first publicly shared social media account documenting my less trashy lifestyle was of a typical pile of groceries. The packaging was excessive, and through my new zero-waste lens I could see that more clearly than ever before. How did I not notice for so many years that I was bringing so much trash home from each shopping trip? Consumers don't buy products for the packaging, so I christened my social media account by documenting the "before" photo in my zero-waste journey, which I felt determined to improve over time as I became less trashy.

Creating this public account would change my life. I had no idea at the time that sharing one aspect of my life on social media was going to be so impactful, nor that it would lead me down the path to becoming an entrepreneur. When I started, I had no followers and everything to learn. While you by no means have to create a blog, I do recommend engaging online to find like-minded people. Online communities can be especially great if you're the kind of person who prefers the company of your pet over people. I like a bit of both, and in my experience, my online community has enhanced my IRL community, and vice versa. Here are a few ways to start developing your online community to make internet (and possibly even IRL) besties:

Join a Facebook group. If you already have a Facebook account, it'll just take a couple of quick searches and a few taps of the mouse and, just like magic, you'll find groups for various topics that speak to your heart. If you're keen to one day meet up with the people you chat with online, try finding local groups. If you want to keep the conversation online, your searches don't need to be location-specific at all. Or join both! It's great to get home-grown perspectives on topics like low-waste living (for example, you might discover a nearby rooftop garden that will take your veggie scraps for its compost bin or learn about a clothing drive for homeless people who need warm coats in winter). It's also worthwhile knowing what's going on globally to learn from others, to see how what you're doing is applicable in the worldwide context, and to learn about how our actions impact the rest of the world (like how our donated clothes end up in secondhand markets throughout Africa). Here are some examples of groups you could look for and how to find them:

- **Buy Nothing groups:** These groups are part of a larger organization, but they have local chapters that you can join. The website for the organization itself is www.buynothingproject.org.
- **Use keywords:** If you're searching for groups to join on Facebook, then use the keywords that interest you. For example, you could include *zero waste* followed by the city you live in, like Toronto. There's a group on Facebook called Zero Waste Toronto. Other keywords you might use could include *low waste, low impact, sustainable, trade, swap, free stuff,* and *repair café.* Basically, just type your own interests into the search bar and see what you find!

Create a public social media account dedicated to your new lifestyle. I started with an Instagram account. I've enjoyed that platform because I can see what other people are doing to live more sustainably and learn from them, while also sharing my experiences. Being a visual learner, I find that the photos and videos make it easy to be inspired and to replicate the tips people are sharing. This platform is where I've made most of my IRL friends. We started by commenting on each others' posts and then moved chatting into DMs. If you're up for it, start your own public account! Just because it's public doesn't mean you need to reveal your identity. You can remain anonymous, if you prefer. I didn't share my full name or my face on my account for almost an entire year. A public account can still be private; ultimately you decide how much to let people on the internet into your personal life.

Read blogs and even start your own. Before I started my own public social media account, and long before I started a blog (*The Zero Waste Collective*), I was reading other people's blogs! There's so much great information online (although, yes, there's also a lot of crap out there—it's always good to fact-check sources). It's a great way to learn about what people are doing to live more sustainably and get inspiration from different cultures, lifestyles, and locations. Read the comments, leave comments, and then, if you're keen, start your own blog or guest write on other people's blogs. The community part comes with it over time. Treehugger and mindbodygreen (look under the "planet" category) are two larger blogs to check out to get you started, and eventually you'll find smaller blogs that might be more local or tied to your personal interests.

Reach out to experts and find mentors. Whether it's field experts (like scientists or journalists) or bloggers on a specialized topic, reach out to them to grow your community and your knowledge. You'll find these experts by watching documentaries, reading the news and other publications, attending conferences, seminars, and networking events, and even going down the rabbit hole on social media, watching You-Tube, or searching waste- and sustainability-related issues on Google. You'd be surprised by how many people respond warmly to an invitation to chat about what they do or share some of their knowledge! You might also collaborate with your city and arrange for a leader you admire to speak at a digital community event, or ask questions during an online conference. Some of my mentors have become part of my network and even friends!

I still remember vividly the first time I met a group of my online "zero-waster" friends *in real life*. There were probably twenty of us meeting up at a restaurant in downtown Toronto. It was a long trek to get to Toronto from Guelph, where I live; I left work early, drove to the train station, jumped on the double-decker train that would take me into the big city, rode the subway to a major intersection near the restaurant, and then walked the rest of the way. The whole journey probably took an hour and a half, one way, but I made it there in time for dinner! Nervous and excited, I recognized everyone from their Instagram photos but, funnily enough, couldn't remember their actual names. I knew them mostly by their handles! I wasn't alone in that; it took everyone some time to actually learn people's real names. I chatted with them eagerly between bites of my portobello mushroom burger, and we all laughed about meeting online as if it were a massive blind date for everyone. And it was! We were all swiping right on how awesome that night was, and I'm still in touch with most people who were there that day.

Make Friends IRL

When I get to the till at one of my regular low-waste grocery stops, I know some of the staff on a first-name basis. "Hi, Ryan! Would you mind weighing my jar today, please? Thanks so much!" It makes my shopping experience that much more pleasant, because it's nice to catch up, say hello, and get the scoop on whether or not more people are coming to use the refill products they have in stock. To me, simply getting to know the employees at my local and independent shops in town is part of growing my green community.

It's also through that network that I've had the opportunity to find out about fun events and to share local information on my social media account. Since changing my lifestyle, I've learned more about my city's waste-management system, because I sit on a public-liaison committee for waste management. I feel in the loop; the information I gather gives me the chance to help inform others with questions, plus I feel engaged in my own town. It can feel pretty empowering. Here are some ideas to build your in-real-life low-waste community:

Volunteer. Volunteering can be an important part of a low-waste lifestyle, because it gives you the opportunity to contribute directly and see the fruits of your labor. It's also a great way to meet new people. You can volunteer with your local municipality, at nonprofit organizations, or with school clubs, among other things. Volunteering can be a one-time thing, like going to a beach cleanup, helping out at repair cafés, or representing an organization at an event and sharing your knowledge. Volunteering can also be a regularly scheduled commitment, like a monthly committee meeting; regular administrative work at your local tool library, helping people find and sign out items; or helping weekly with a nearby community garden. The options are limitless, and there will always be something to cater to your needs, interests, and schedule. Find a cause that's most important to you and do some research to find out how you can be of service in your community. It's the perfect way to meet new people and have a great time!

Join a club. Join any local clubs at school or in your community that are focused on sustainability. It'll give you a chance to work on cool projects, meet new people, learn a few new things, and expand your sense of community in a productive and helpful way.

Create a green team at work or at school. When I was still working in a small office of about twenty-five people, I started a green team. It doesn't matter how many people you have in your workplace, let everyone know you're starting a green team to enhance it, expand everyone's knowledge, and have a positive impact in the office and on the local community. Just be sure to get the go-ahead, if you need to, from a supervisor first. Then you can send out an email and/or share the news through online platforms and invite people to join. Your green initiative can be formal (with meetings, minute taking, and regularly scheduled events) or informal (casual get-togethers and occasional events). Most important, have fun! While I was really proud of the action my office's team took, my favorite part was eating pizza after a big green team litter cleanup. Is that so wrong?

Shop local. Switching from the big grocery store chains to the small independent businesses where you live is not only a great way to support your local economy but also a wonderful means for improving your sense of community. When you shop local at independent businesses and places like farmers' markets, it's easier to see the same faces on a regular basis and get to know people in your town. It's refreshing and helps weave a sense of belonging, plus it'll give you the opportunity to learn more about what's going on in your community (like sustainability events). Shopping local also helps us reduce waste, because independent stores tend to be more flexible in responding to new trends, like offering refill programs and being supportive of reusable containers and bags. Big-box stores might have more red tape in terms of updating policies to support low-waste shopping.

Participate in events. Attending events is entertaining, but you can often volunteer to help make public events in your community less trashy and more sustainable. Use your passion and interest to come up with ideas for reducing the waste footprint of events like festivals, concerts, outdoor movie nights, art exhibitions, pop-up markets, religious gatherings, and more.

Go to conferences and trade shows. Conferences on sustainable living are great places to learn and network, and sustainability trade shows offer similar opportunities, with a bigger emphasis on promoting an industry's products and services. I love both! I've learned from amazing speakers (and even been a speaker myself!), met new people, and discovered really innovative up-and-coming businesses at various green-living-focused conferences and trade shows.

Attend climate strikes and sustainability marches. Sometimes strikes and marches feel a bit out of my comfort zone (I show up anyway), but they're a great way to get social, spread awareness, learn more, meet new friends, and put pressure on the government for change.

- www.globalclimatestrike.net is a great resource for climate-related demonstrations, and while not directly a waste-related platform, the main issues are deeply interconnected, because waste contributes to climate change. For example, organic material that breaks down in landfills generates methane gas, which is a greenhouse gas contributing to climate change.

Get civically engaged. From voting to attending public meetings, exercise your right to participate as an active community member with your local government council. You'll meet new people who also actively participate in community decision making, from city staff and elected officials to other engaged citizens. You can make your voice heard through official channels when you participate in this capacity. Your local government's website should be the best place to go for information on civic events, committees, and other projects you may want to have your say on. If you can't find what you're looking for, give your city hall a call.

Having worked at both municipal and provincial levels of government, I was amazed by how many people (lots) have strong opinions about their community but can't be bothered to peel themselves off the couch to do anything about them. Complaining is often about as far as they get. I often saw the same few faces, heard the same few voices, and read emails from the same few people. I knew there were a lot of people with strong opinions and smart ideas out there, but I often felt a broad representation of the community's feelings was missing from important conversations around government-related decisions.

I get it; going to council meetings can be a total snooze fest. Policy and legislation are two uninteresting (at least compared to whatever Kim Kardashian is wearing) government tools that enable the levels of greenhouse gas emissions and amount of waste we see today. These same tools could be used in different ways to instead limit and reduce emissions and waste. Having a positive impact on the world isn't always as glamorous as a photo of Leonardo DiCaprio and Greta Thunberg hanging out and chatting about climate change and the current chaotic state of the world. It's important for us to put the necessary pressure on government to make the changes we want to see.

What I'm saying is, show up. If you're keen to show up to climate marches, that's great; those are important, too. But while you're at it, show up and use your voice (share your point of view), your presence (attend meetings), your words (send an email), and your vote (once you are of age) to impact the direction of legislation and policy for a more sustainable and less trashy future!

The way you build your community will depend on your personality type, how much time you have available, and the skills you can offer. There are so many ways to participate actively and to develop online and IRL friends and a strong sense of community in your sustainable-lifestyle journey, and you get to decide how you would like to do that. Having this community can help keep you actively engaged and hold you accountable, while also making everything you do much more fun.

Reflection and Checklist

Your close friends and family members may or may not be on board with your sustainable-lifestyle changes. That's fine; they don't have to be your sounding board for *all things* less trashy. I learned this the hard way one day when Hubby finally said to me, "Can we talk about something other than zero waste?" Of course, he didn't mean forever—he just meant that he wanted to have conversations again that didn't all somehow end up on the topic of sustainability and zero-waste living. He wasn't rude about it; he was just being honest. And fair enough! He shouldn't have been my only outlet for those types of conversations.

He was growing tired; while my husband was on board with making the lifestyle changes in our home, he didn't have the same passion that I did.

Do your best not to tire your friends and family out with these topics if they don't share your level of excitement. Hopefully you'll try out some of the strategies laid out in chapter 8, though, because their support is so crucial. And if they don't share any enthusiasm or interest, then you'll definitely need to find some like-minded friends online and in real life. Here's a recap of some simple steps you can take to grow your community and give your friends and family a break from your newfound obsession about living less trashy:

- Make new friends online and IRL through social media
- Start your own public social media account or blog documenting your experience and knowledge
- Connect with experts and mentors
- Volunteer in your community
- Support local businesses
- Go to and help organize events
- Become an engaged citizen with your government

Chapter 10

Money Matters

Bust the myth that zero-waste living will break the bank

It's Taboo

The topic of money was always a bit too taboo in my house growing up. My dad was the primary income earner, through his work as a pilot, while my mom put in the work at home raising me and my sister. I had a unique and exciting upbringing, living the expat life for many years in the Middle East. My parents both came from working-class families, and before my dad landed his job in the late 1980s, they didn't have much money. I was too young to remember what our lives were like before moving to the Middle East, when we lived modestly in a relatively remote and northernish location in Canada, where I was born. It's a tiny town called Timmins, and its only claim to fame that I know of is that it's the hometown of the country singer Shania Twain.

It was off-limits to discuss my dad's salary, but from what I can remember, lack of money didn't seem to be an issue. There were always plentiful gifts and lots of food during holidays, and my parents paid my way through university (twice), including (*gasp!*) my rent and groceries. I'll be honest with you; sharing that makes me feel incredibly uncomfortable, because I know that's not most people's experience and it's very privileged. These days most graduates leave school with mountains of student debt. Having both come from humble backgrounds, my parents are still over the moon to have two daughters who graduated from university. They didn't go to university, and neither did their parents, and because of that, our university education brings them a lot of pride. I'm extremely grateful that they gave me the opportunities they did, and I will always be thankful for starting my working life debt free.

The problem was, though, that I didn't learn how to manage money. What I learned was how to spend it! Yes, there was the occasional conversation about how I should save for retirement, but let's get real: What teenager, or even early-twentysomething, cares about saving for retirement when that feels like forever away? For me, at least, it wasn't exactly my biggest priority. Surely my parents had conversations with me about finances and saving, but I just don't really remember them. Instead, I was interested in shopping and spending money, and my financial lifeline was my parents.

While I did work as soon as I could, babysitting at fourteen and landing a fancy grocery store job at sixteen, making salads and chopping fruit, I spent that money, too. I spent money with such great success that my gap-year trip to New Zealand between high school and university was cut short by five months because I went broke twice. The first time I went penniless on that trip I got topped up from the Bank of Mom and Dad, and the second time I came home with no money to my name. Was it *really* my fault? I mean, New Zealand is the land of fun, with endless activities, from bungee jumping and skydiving to sailing trips and the *occasional* party. I had a work visa in New Zealand, and even with my odd jobs I still couldn't figure out how to balance my income and expenses to make my trip last the full year.

Though I am grateful that I didn't have to worry about money growing up or in school, money still stressed me out because it always seemed to run out. What was the real issue here? To sum it up, I didn't know how to handle it properly. Just like the government, I couldn't balance the budget.

It wasn't until I moved in with my boyfriend, now my husband, that I discovered our financial habits were totally different. I was living it up like a baller (without the millions of dollars), and meanwhile Hubby was diligently budgeting, saving the money he earned, and paying off any debt he had as soon as possible. There are very few people who graduate from university without student loans as a result of paying their own way (maybe with a bit of help from their parents), but he's one of those miraculous and rare people (a unicorn, basically). What a responsible guy! And also kind of annoying when you're a spendy-pants like me.

Hubby grew up in a household where his needs were met in a comfortable and loving home, but if he wanted any extras, he had to get a job and pay for them himself, which he did. For example, his parents would give him fifty dollars toward a pair of shoes (a need), but if he wanted brand-name shoes that were more expensive (a want), he'd have to make up the difference; these were financial lessons he learned from his parents. As a result, he's essentially a money master in my eyes. It's safe to say that our childhood and teenage experiences with moolah were part of the reason we came to argue about how much we spent on groceries.

Yes, *groceries*. Of all things, who knew our biggest disagreement would be about how much to spend on food each month? He was a bit of a Debbie Downer (sorry, sweetie, I love you!) when it came to spending money on fine cheeses and nice crackers; Hubby was a coupon clipper all the way. It took me some time to adjust to our new grocery budget, but what I didn't realize at the time was that this new way of life was setting the foundation for a much more financially responsible Tara. I'm way more into saving and investing now than I ever was before, and I believe that's in part because of maturity but also because I learned lots from Hubby. Recognizing this knowledge gap, I've since undertaken my own self-education journey on finances—money mastery, here I come!

I'll finally get to the point here. I've been blogging about low-waste living and sustainability long enough to know that *loads* of people have the perception that the trashless lifestyle is hella expensive and financially out of reach. This limiting belief is so widespread that people seem to put off adopting the lifestyle because they think it'll cost money. I have to dispel that myth now: A low-waste lifestyle *doesn't* have to be expensive. It *can* be expensive; it's just up to *you* whether it will be.

You can choose to make your sustainable lifestyle cost less instead of more. How? You simply have to become aware that you have choices and learn what your options are, so that you choose the most kick-ass path forward. Like, not everyone needs to buy reusable metal straws to be sustainable. Save your cash and skip straws altogether! If you can, that is. It's become well known that plastic straws are necessary in certain situations for accessibility purposes, and if that's the case for you, please use plastic straws.

When you let some (or all) of the lessons from this book guide your less trashy lifestyle, you should be spending far less moolah than you did before. Unless, of course, you're a spendy-pants, in which case, this month is exactly what you might need to take charge of your not-so-thrifty habits. Either way, here's the lowdown on what's ahead: find ways to be frugal; save for the big-ticket items; and put your money where your mouth is. This month is kind of like a mental boot camp for your conscious money mindset. Ready, set, go!

Find Ways to Be Frugal

There are lovely photographs on social media, in blog posts, and in the news showing off beautiful zero-waste kits full of stainless-steel straws, bamboo cutlery, insulated and stylish coffee cups and gorgeous water bottles, cute organic-cotton net bags, and matching sets of glass mason jars. You might even find these pretty photos on my website and social media (okay, you *will* find these photos). These items are useful, and if that's exactly the kind of kit you want to start your low-waste lifestyle, then go for it! If products like these can convert those who are addicted to single-use plastics, then I'm all for it. Plus, I'll admit that I personally like an aesthetically pleasing reusable water bottle. But a full kit with all the shiny bells and whistles can cost a pretty penny.

The thing is, just because these products have been popularized by the zero-waste movement doesn't mean you have to buy them! There's no initiation into the lifestyle requiring you to buy a bamboo cutlery set for eating on the go and matching organic-cotton tote bags for your trips to the local farmers' market. This is not what reducing your waste is about! Yes, having all the reusables is great. They just don't have to be shiny and new, perfect or pretty. Instead, use what you have first. If you find you still need (or want) a few items to help you put together your ideal zero-waste kit, then you can likely find ways to create it while still being economical if you're budget conscious: look for free stuff, shop secondhand, swap and trade. Here are some things you can include in your zero-waste kit that you probably already have at home:

- Cutlery
- Napkin
- Reusable water bottle
- Tote bag
- Travel mug
- Tupperware

Living less trashy is about more than a zero-waste kit, too. It's about being a conscious consumer, which is a concept we explored in chapter 3. It's also about consuming less in general, which saves resources and money. While it's great if you can bring your own bags and containers to the local grocery and bulk shops, not everyone has access to this style of shopping, and there are so many other ways to reduce your waste that don't involve shopping.

That's actually the key: Don't shop! Here's the alternative: Simply shop less, and when you do shop, choose better and fewer things that will last longer and can be repaired over time. You can still find high-quality items that are free or cost less secondhand. Basically, anything you buy brand-new will lose its value as soon as you take it home (buying a brand-new car is a great example of this immediate depreciation). That means if anyone is reselling something, it will automatically be cheaper than buying it brand spanking new. When I bought a car shortly after starting my first full-time, career-oriented job, I was ready to drop $30K on a fancy new one. My husband, the money master, reminded me that it was a lot of money to have to repay, plus the car would lose value. While I definitely was tempted to pick all of the options (like heated seats) that come with a new car, instead I found a year-old version of the Ford Fusion I wanted and spent $20K instead of $30K! It didn't come with heated seats, sadly, but the following Christmas my husband had that feature added to the driver's seat of my preloved car so my bum will be warm every winter! I still have that car eight years later. I'm grateful I didn't spend that extra $10K; I paid it off quicker, and now I'm debt free in the car department.

Instead of buying at all, you can also rent or borrow what you need (cars included). Another option, which is slightly countercultural, is to not buy anything at all that isn't a necessity (e.g., food and soap). We'll explore this idea further in the next chapter: "No-Buy Month." You'll discover that an eco-friendly lifestyle doesn't have to cost an arm and a leg. For now, here's a quick recap of ways that you can be frugal with your less trashy lifestyle (which are recurring themes throughout the book):

- Use what you have (like items for your zero-waste kit)
- Repair instead of tossing and buying new (like the toaster oven you thought was fried)
- Find free stuff (like furniture)
- Borrow instead of buying (like specialty tools)
- Rent instead of owning (like carpet cleaners)
- Swap and trade (like toys)
- Shop secondhand (like clothes)
- Choose quality over quantity (like everything!)

Save for the Big-Ticket Items

We live in an era of instant gratification. Gone are the days when people would save up for weeks, months, or even years to obtain the items they wanted to buy. This is what our grandparents would do; they didn't have a choice. Instead, we're now living a life on credit, shopping online and getting everything our heart desires delivered to our door the next day. You can even get payment plans to buy clothes! It's nuts. The problem? It's making us broke AF, and we're paying interest on our interest, and the debt keeps piling up.

Here's a simple rule to follow: If you don't have the cash for it now, don't buy it on credit unless necessary—being in a tight spot happens. Use a credit card to build up your credit score, and make sure you can pay off your credit card each month, as well as all your other bills. If you know you can't pay it off by the deadline, then save for the item (small or big) or trip (local or afar) that you want. It's not as fun, and it's unlikely that there will be anyone around to pat you on the back when you do this (I suggest patting yourself on the back, because it's a kick-ass accomplishment), but it'll be a lot less stressful.

Big-ticket items can seem like a major challenge to save for, and they might feel out of reach. When Hubby and I got engaged, we knew our wedding was going to be one of those big-ticket items. We went all out for it, with 130 guests, a three-course dinner, and an open bar. The best part of it all? We budgeted for it from (almost) day one. When we landed on the date of our wedding, which was eighteen months out from our engagement, we worked backward from that date to budget. While both of our parents kindly and generously contributed to the bill, we still had plenty to save for. This approach enabled us to save the total amount we needed from each paycheck to pay off the wedding and come out debt free.

Hubby and I even opened a special joint bank account to make sure we were saving and not spending the money. Was it my idea? Of course not! Spendy-pants over here (me) was more like, *Ughhhh, do we have to?* Obviously, I'm glad we did. It took planning and discipline, but it worked. It just takes one step at a time. Here's how you can do the same thing to save for your special big-ticket items:

Find out the total cost. Research the item you're buying or the trip you're taking, and find out how much it will cost. Factor in all costs involved, especially if it is an item that requires accessories or insurance. In the case of a trip, there will likely be many costs to factor in, such as transportation, accommodations, and food.

Decide on a date to reach your goal. Whether it's the day you want to buy a new fridge or to leave on a big trip to Morocco, pick a date, so that you know how much time you have between now and then to save for the total cost.

Do the math. Divide the total cost by the number of paychecks you have between now and the date you landed on above, and you'll get the amount you'll need to save from each paycheck.

Make sure it's in the budget. Once you do the math, make sure it's in your budget. Say your savings goal is one hundred dollars, and you need that amount in four weeks. In the next four weeks you know that you'll get two paychecks, so you'll need to save fifty dollars from each paycheck to make your savings goal. If you can't afford to save fifty dollars from each paycheck, consider revisiting either the total cost (maybe you can change the item or something in your trip) or the deadline (move the date of purchase or your trip date).

Get a handle on your finances. If all of the above seems overwhelming, or if your current debt seems out of control, then it's time to get back to basics. There's a lot of valuable *and free* information at the library and online that can help you get your finances on track! If need be, talk to friends and family who are good with money or seek professional advice. Just as it takes training to become an Olympic athlete, it takes training to be good with money, too. I truly believe that giving yourself a financial education will pay off huge for your future *and* can give you the opportunity to invest more in the planet (more on that below).

Create an emergency fund. If you keep your credit card handy for emergencies (the unplanned big-ticket items), consider saving up at least a thousand dollars that you don't touch except in emergencies, so that you don't always rely on credit. Similar to the technique above, just take it step by step to save what you can to make your goal of one thousand dollars, which is the amount recommended by financial guru Dave Ramsey. It's even better if you have more tucked away, but a thousand dollars is a great start!

Saving for big-ticket items is an old-school approach that needs to come back into style; it'll help us reduce our consumption and improve our finances. Credit card companies don't want us to figure this out. And why is this useful for sustainable living? Well, I often recommend choosing quality over quantity, and quality products can be more expensive because they are made better, built to last, and often (but not always) more ethically and sustainably produced.

Saving is a skill. Money management is an aptitude. Unfortunately, these tools aren't always taught in schools or at home. Flex those financial muscles by learning more, creating a budget, saving, and investing! It'll help you get off the consumer treadmill and get you on top of your finances. You'll eventually get to a point where you'll be out shopping with your friends and say, like a boss, with a well-timed hair flick, "Please, I'm putting my money in my retirement account instead of buying that two-hundred-dollar gold sequin dress I'll only wear once!" Yes. Consume less, save money, and help the planet. That's winning all around!

Put Your Money Where Your Mouth Is

I've always been inspired by the story of the late Doug Tompkins, billionaire and a founder of The North Face, an outdoor equipment and apparel company. As a wilderness lover, he bought massive tracts of land in Chile specifically for nature conservation. A whopping 408,000 hectares (1 million acres) of land has since been donated by his wife, Kristine McDivitt Tompkins, to the Chilean government to create a network of seventeen national parks, providing both education and employment.[1] Kris continues this legacy of environmental leadership through their nonprofit organization, Tompkins Conservation.[2] Talk about putting their money where their mouths are!

There are many ways we can actively advocate for the environment without money. We can educate those around us about the issues; attend climate marches; reduce our consumption and waste; vote for political leaders with strong environmental agendas; participate in civic affairs; write to corporations; and more. Yet we have limited control over the outcomes of these activities. For example, we can vote for political leaders who represent our values, but that doesn't mean they'll be elected. What we do have control over is our own actions. We can use our money to vote for the world we want to live in. I think that's a great approach for things like buying ethical and sustainable products, yet there's so much more we can be doing that can have direct impacts, like donating money to support vital causes such as nature conservation. I see it as investing in the planet's future.

While not all of us are billionaires like Doug and Kris Tompkins, we can still financially support important environmental causes. You can start today! Pick a project or an organization that you want to support and donate some of your moolah; better yet, make it a monthly habit. As your income grows, you can donate more! You'll be helping save the planet directly, even if it's ten dollars at a time.

Reflection and Checklist

Living less trashy is about reducing your waste in the context of your budget, what's available to you, your personal circumstances, and your lifestyle goals. It's not about keeping up with the zero-waste Joneses. It doesn't have to cost a lot of money. You don't have to buy all the fancy reusable products or shop at specialty food stores, unless, of course, you want to. You know yourself best, so make the judgment call with that in mind.

You can be frugal, and you can use your money to support important causes, as Doug and Kris Tompkins have. Don't be afraid to donate what you can. Whether you have ten dollars or ten thousand dollars to contribute, it all matters. Here's a recap of this chapter's conscious money mindset boot camp:

- Find ways to be frugal with your less trashy lifestyle by using what you have first, trading and swapping with friends and your community, buying secondhand, and looking for good quality
- Save for your big-ticket items; don't get caught up in the hamster wheel of consumption by seeking instant gratification from immediate purchases on credit; while you're at it, invest some time in your financial education
- Put your money where your mouth is by supporting the environmental causes that you're most passionate about; if your budget is super tight, start with what you can, even if it's five or ten dollars

Chapter 11

No-Buy Month

It's more doable than you think

You Only Live Once

I've been a big spender for most of my life. If someone had challenged me ten years ago to spend no cash, except on necessities, *and* to keep that behavior up for an *entire* month, I'm sure I'd have rolled on the floor laughing . . . then said no. After all, I had been an avid and committed consumer who spent my "extra" money on clothes, eating out, going to the movies, concerts, nights out at the bar with friends, and a gym membership. Why on earth would I want to give all that up? Habits are hard enough to break when you *want* to make a change, so I can't imagine how difficult it would have been back then, when I was humming along, basking in the status quo of spend, spend, spend. Of course, my thoughts were more like *YOLO—you only live once!* My spending was justified because it meant I was living my best life!

Fast-forward to the new, more financially responsible Tara. Doing a no-buy month has become a no-brainer! In recent years I've been way more cautious about my spending, but I still spent more than I wanted to in the lead-up to the 2019 holiday season, so I decided to do my first no-buy month in January 2020. I had read about other people doing no-buy challenges in various blogs on minimalism that I had been reading but was mostly inspired by Cait Flanders's book *The Year of Less,* where Cait documented her year of not shopping. A whole *year*! I'm only asking that you try a no-buy *month*! Anyway, a no-buy month is a challenge where you avoid shopping in specific categories of your choice (like clothes or home decor) for a whole month. I was inspired by the fact that Cait managed to pay off nearly $30K of debt in the process. What strikes me so much about it is that it goes to show us that we can rein in our consumption habits to live with less, which also results in less waste (as we explored more closely in chapter 2 on minimalism).

I was feeling a bit financially hungover, and January seemed like the best time to try to quit cold turkey in the spending department. Cold turkey except for minor but important exceptions, like, you know, paying the mortgage and buying groceries. I was inspired to undertake this challenge to save money and reduce my consumption, which in turn would also reduce my waste. This, I knew, would be a new and different way for me to document my consumption patterns to help me live with less beyond the no-buy month. It's taking minimalism to the next level!

A no-buy month gives us the chance to slow down, get off the consumer hamster wheel, and undertake a self-assessment. This is an opportunity to come out the other side with a clean slate where we call the shots (like taking more time to make purchases or choosing not to make them at all) rather than succumbing to societal norms and marketing. It's similarly a chance to become more resourceful and less materialistic, focusing on other things in our lives that we enjoy, like spending time with loved ones, rather than online shopping.

I was hoping that this challenge would be empowering and not a complete drag. Were there hard moments? Sure! January tends to be a slow month for the sales world because people are still paying off their credit cards from the holiday season. As a result, you'll often find things on sale, and it's tempting to shop, but I held my ground. You'll probably experience something similar; an event may come up, a special sale will pop up, or your friends might ask you to go out for drinks or a shopping date. Those are going to be hard moments to say no, but keep in mind the benefits and why you're doing it in the first place.

Before you set out to plan your own no-buy month, it's important to remember that it isn't about restricting yourself. It's more about enabling yourself to take control of your life rather than living on autopilot. If you're wondering if it's worth your time, think about how quickly a single month goes by. From there, if you like the changes you've made, you can make it an annual trend or tack on a few more months—maybe even try it out for a whole year, if you're feeling like you really want to commit to a big change!

Paying off debt and saving for a big-ticket item like the down payment on a house are both perfect reasons to do a whole no-buy year. I recommend starting with a no-buy month, because research shows that it takes a minimum of three weeks to develop a new habit. A full month is the perfect amount of time to let this challenge start changing your life for the better, but you can start even smaller, with a no-buy week, if you want to ease yourself into it. As your habits begin to shift and you feel more comfortable with the changes you're making, it might be the right time to think about extending the challenge—if that's what you want to do. When it comes to planning and actually doing the challenge, we'll keep it simple. Here's what you're in for: define your motivation, set the ground rules, go for it and journal it, and learn from it. That's all!

Define Your Motivation

The previous chapter, "Money Matters," hopefully warmed up your financial muscles in preparation for this challenge. The reflection and review of your current moolah situation should have you revved up to rein in your spending. Ultimately, whether or not you're successful in completing the challenge comes down to your personal motivation to stick with it. When I think back to my first no-buy month, I was 100 percent into it. That's because I was very clear about the reasons why I was doing the challenge in the first place.

Before you actually decide what you're going to cross off your shopping list for the entire month, take the time to sit down and ask yourself why you're making the change. I decided to do the challenge as a way to evaluate my consumption, reduce my waste, explore minimalism further, and become a more conscious consumer. You may have similar inclinations, but you get to decide for yourself what your motivation is! There's no right answer or specific number of reasons that you should have before embarking on this journey. Consider reflecting on and, if you want to, writing down responses to questions like these:

- Why do I want to do a no-buy month (or week or year)?
- What specific outcomes am I hoping to achieve? (To pay off a certain amount of debt, build a new habit, save for a big-ticket item, etc.)
- Are there other related intentions I have that this challenge will help with? (Maybe you want to stop eating out, explore minimalism, learn about finances, work on decluttering, etc.) Here were some of my intentions going into my no-buy month:
  - Enjoy and use what I have
  - Explore minimalism a little more deeply

- Determine what I really need
- Save money and develop better budgeting and spending habits

Set the Ground Rules

The nice thing about a self-imposed no-buy month was that I got to make the rules. That's probably why it felt so reasonable to do the challenge, and why I didn't feel like I was depriving myself for the month. You don't have to deprive yourself either, but some ground rules will help make it effective.

Start by getting a snapshot of your current financial situation. Maybe you reviewed your finances after reading the last chapter, but if you didn't, now is the perfect time. Whether you go all out and create a detailed spreadsheet of your income and expenses for the last few months or you choose to just briefly glance over your most frequent (and unnecessary) purchases is up to you. But a word to the wise: The more specific you are in knowing where you spend your money, the better. Once you have an idea of where your money goes, you can determine more easily the types of things you want to stop spending money on during your no-buy month.

Start by making a list of necessary expenses that you will continue to incur (like rent/mortgage, bill payments, food, toiletries). Then make a list of ground rules, including what's going to be off-limits (like clothes, shoes, makeup, home decor, tools, hobby equipment, etc.) and what isn't. My list looked something like this:

- No buying stuff (clothes, books, decor, or other household things)
- Okay to spend on necessities like household bills, food, toiletries, business expenses
- I will still buy experiences and eat out occasionally

- Find free alternatives, like:
  - Going to the library and tool library as needed
  - Doing activities and experiences that are free (like hiking and game nights)
  - Hosting potlucks and get-togethers

Based on my self-inflicted (I mean, developed) ground rules, it was a successful month! I didn't do any shopping (other than groceries) and ate out only a few times. While I could have removed eating out from my list of exceptions, I chose not to because my main purpose was to stop purchasing tangible "stuff" for the month, rather than to 100 percent limit my experiences. Still, I decided to do far fewer paid activities, while focusing as much as possible on free events and hobbies, so I went on more hikes in the woods and had more potlucks with friends.

Feel free to use this to develop your own list of ground rules. Or you can make your own completely from scratch! Now you need to set your start and end dates for the challenge. Once you get started, be sure to refer to your ground rules if you get tempted to buy something.

Go for It and Journal It

The best thing you can do for yourself during this no-buy month is to fully follow through on your ground rules. While you're at it, journal about it regularly (daily could be helpful), document it on social media, and/or tell your friends and family about it. Whether publicly or privately, documenting your progress will help keep you accountable to yourself. If you decide to share your progress online, you may also find that the positive reinforcement you receive from your friends and family helps keep you feeling inspired.

Learn from It

What was your experience like during your no-buy month? Set aside some time to reflect on your initial reasons for taking on the challenge, determine whether or not you met your goals, consider if you learned anything you weren't expecting to, and decide if it provided an opportunity to explore other curiosities (like learning more about finances, if that was on your list). Write down what you took away from the experiment, or at least give these things some thought, and maybe even share your experiences online and with friends and family. What did you learn? Would you do it again? Did you break any of the rules, and if so, why? It's these lovely life lessons that'll get your ass in gear moving forward. Whatever you do, don't make up for not shopping by binge-shopping afterward!

If you're like me, you may discover by the end of the month that your mindset has shifted. For me, it became very clear that I don't need to buy everything that I want. Like me, you'll probably learn to better decipher what you *need* versus what you *want,* and then realize that you don't need as much as you previously thought. That's the game changer! When your perception shifts and you start to see the world differently, you can carry your new outlook with you into the future, even if you decide not to maintain the strict parameters you set out for yourself during your no-buy month.

If you do decide you're going to go all in on this lifestyle choice and commit to a no-buy year, good for you! But even if you choose to let the challenge go completely, or to repeat it just once a year, you'll find that the monthlong reset will help set you on a less consumer-driven trajectory. In the long run, it's what you learn about yourself through this experiment that is most important, because you'll take it with you forever.

Since completing my first no-buy month, I've spent very little money on stuff that I don't need. And when I did make a few purchases in the following year, I made sure I took time to determine whether I would get enough use out of each item to justify the purchase. Essentially, I developed the perspective of a badass conscious consumer.

I've also found that shopping less results in less of what I call "stuff management." Stuff management is all the work you have to do to bring a new item into your home. So if you order something online, for example, the package arrives at your house, and first you have to deal with the packaging. Then you have to either wash the item, if it's new clothing, or find a home for it. Beyond that, stuff management is also about the care and maintenance of an item over its lifetime. Stuff management is making sure things are cleaned, maintained, and put away where they belong. The more things we own, the more stuff management we have to do. I learned that I prefer less stuff management, which also drives my personal preference to own less and shop less. Looking back, I'd say it was "mission accomplished" in terms of my stated goal to explore minimalism a little more deeply during my one-month no-buy challenge.

Minimalism and working toward more sustainable living through reducing my waste both put me in a good place to do this trial. I was ready to kick ass, and I've changed for the better as a result. I'm way more careful with how, where, and when I spend my moolah (meaning that I spend it less), and I no longer feel part of the consumer rat race. I still like to buy stuff, of course; I just don't do it as often.

Reflection and Checklist

After completing the no-buy challenge, I felt as though I had more control over my life, which was an odd feeling, because it wasn't like I had any less control before I did the challenge. And I still felt like I could live my best life! It's not as if when I cut off excess expenditures my life suddenly became bleak, boring, and bankrupt of excitement. The difference was that I became more aware of my spending, and I saw the success in my bank accounts. And that felt so, so satisfying. With my decreased spending, I got my credit card balance down to zero. It gave me such a feeling of freedom. Once you start to see the positive changes on your financial statements, saving can become just as addictive as spending, or even more so! The trick is to get to that point, then to stay there. A no-buy month is the perfect way to start.

Talking about money can be cringeworthy for most people, but plugging our ears and saying "lalalalala" while trying to ignore the debt we're collecting will just keep us down. Instead, an activity like this no-buy month will get your butt in gear, get you off the hamster wheel of thoughtless spending, and allow you to feel way more in charge of your life. Here's a summary of this month's activities to help you kick ass in your finances and reduce your consumption:

- Know your "why"; decide what your motivation is for committing to a no-buy month
- Outline your goals and any other related intentions you have for the challenge
- Write the rules for your no-buy month by determining which purchases are essential and which ones are not
- Pick your dates and just do it!
- Journal about your progress, talk about it, and share with family and friends and on social media to keep yourself accountable
- Reflect on and learn from your experiences; take those life lessons with you moving forward

Chapter 12

Wrap It Up

Define your legacy

That Pivotal Moment

Whenever it gets tough, when I want to give up on sustainability and throw in the towel because it seems too hard or like no one else cares or that the world's on fire anyway, I always have to go back to that pivotal moment in my life years ago. The moment I lost my best friend to a tragic accident, which happened within the same twenty-four-hour period when I was so struck by the environmental devastation and plastic waste I witnessed while snorkeling in Bali. While my extreme grief has since passed, I'm still saddened when I reflect on both events. Over the years, both of these experiences have fueled my journey in powerful ways.

My best friend's life was cut too short at the young age of twenty-six, and I'm still here, so not only do I have an opportunity to spread the word about living lighter on this planet, but I also believe it's my obligation. It's my responsibility to live my life to the fullest, and to have the most positive impact on the people and planet around me, because I am here. While she died young, my best friend left a legacy behind; her presence created a sense of peace, kindness, and love that radiated from within her, and being around her brought out these characteristics in other people. I believe that the memory of her enables the people she knew to continue to live, laugh, and love to the fullest. I know I won't be here forever, but I hope the legacy of the work I'm doing will have a ripple effect on the words and actions of others. That's the legacy I want to leave behind, and I think it's a worthy one for all of us.

That got deep fast! The point is, life isn't always rainbows and butterflies. It can get tough. We may want to give up. And the first thing to go, quite often, is our sustainable-lifestyle choices. We don't have to be perfect, though. I have to reiterate that incredibly important point, because if you put this lifestyle on a pedestal, it may feel out of reach. Remember this: You do not have to aim to fit all of your trash from an entire year or month or even week in a mason jar. That may be the path forward for some, but it won't work for everyone. I personally can't live up to that expectation. Being less trashy is about making better choices and building the habits necessary to make it easier to live with less waste.

This final chapter wraps up your Don't Be Trashy Challenge and marks the start of your lifelong low-waste journey. This chapter gives you the chance to pull together everything you've learned, tried, and accomplished up to this point. What's most important now is what you choose to take with you moving forward. It's the habits you'll decide to develop and keep, your definition of what sustainable living means to you, and finding a life with more joy in the process. There's no cookie-cutter solution. You get to make the rules! You do you, in the following steps: reflect on what you've accomplished since you started this journey to become less trashy, and develop habits based on the low-waste lifestyle you want to live. Setting your own standards will be much more effective than trying to keep up with all the tips, suggestions, and lifestyle changes I've laid out. I hope they prove helpful, so if you like them, use them and put your own mark on them! Take what you love from this book and leave the rest.

A Time for Reflection

Reflect on—and perhaps journal about—what you've learned, what you liked, and what you didn't like about the transitions you've made since you started reading this book, and then decide what you'll take with you moving forward. Whether you realize it or not, you've probably made so much progress! This is also a good time to define what sustainability means to you. Some sustainable-lifestyle actions will be loved by some and loathed by others. For example, I am *not* a do-it-yourself kind of person, like, *at all*. If I can buy nicely made beeswax wraps and all-natural, nontoxic deodorant in plastic-free packaging, then I'm not going to spend my time trying to make these for myself. But this is my *personal perspective and preference*. I know so many people *love* DIYs! And often they can be more budget-friendly than their premade counterparts. Basically, you just have to tease out what you like and what you want to do, so that you stick with it! Here's a quick recap of the concepts explored over the past eleven chapters:

1. Less trashy lifestyle 101
2. Decluttering and minimalism
3. Conscious consumption
4. Kitchen, food waste, and grocery shopping
5. Bathroom, beauty, and cleaning
6. Sustainable fashion
7. The art of refusal and saying no
8. Family and friends
9. Finding community

10. Budgeting and money
11. No-buy month

With these in mind, it's time to reflect on what topics, information, tips, and lifestyle changes were most appealing to you. Grab a notebook, open up your Notes app or a Word doc, or take mental notes, and answer the following questions:

- What are my favorite topics?
- What do I want to learn more about?
- What habits am I keen to develop?
- Which lifestyle changes am I likely to make and keep?
- Which lifestyle changes should I not bother with, because I know I won't follow through?
- What does sustainability mean to me?
- What are my lifestyle priorities?
- What does less trashy living look like in my life?
- What does it mean to me to live a life with more joy?
- What legacy do I want to leave?

Living less trashy can sometimes feel at odds with some of the ideas explored in this book. For example, I love minimalism. From a clean and uncluttered aesthetic to consuming less, it just makes so much sense to me. Its ethos resonates with me and seems intrinsically sustainable from the less-is-more approach to that lifestyle. However, it starts to get muddled for me when I want to keep stuff that I don't really need but I don't want to waste. Hubby falls into this trap, too. Unfortunately, our garage became the number one place where odds and ends went into "storage" for later, aka the stuff graveyard, where odds and ends really went to die. We finally got around to cleaning it out, and we found new homes for as many of the discards as possible. A lot of it was scrap pieces from building and renovation projects Hubby worked on, including various pieces of wood, wiring, drywall, and scrap metal. We rehomed as much of it as possible by giving it away for free and taking metal to scrap bins. Finally, the rest went in the trash (sad face).

This type of "collecting" can happen with absolutely anything you own, once you decide you want to be less trashy. There will be times when minimalism will be at odds with a low-waste lifestyle, but remember, if you know you won't use it, and you definitely don't love it, then it's just trash sitting in your own home. You don't want your home to end up on an episode of *Hoarders;* that was for sure the fate of our garage if we hadn't finally cleaned it out.

That last example was a roundabout way to say, set your own priorities and decide what's most important to you. If saving odds and ends because you know you'll use them is important to you, then do it. If you know you'll be better off letting stuff go, then do it. Similarly, traveling (especially flying) can be at odds with sustainable and less trashy living, too. Instead of giving up on either travel or sustainable living altogether, find ways to make it more mindful and less impactful. You make the rules, and you get to decide what sustainable living looks like in your life. Now's the time to define what that means. Go forth and reflect!

Develop Habits Based on the Lifestyle You Want to Live

Do you always forget your reusable bags at home? Or do you leave trip planning to the last minute, so you end up driving when you could have taken the train? Or do you go for coffee but forget a reusable mug? If these situations sound like your life, then it's time to build new habits that make sustainable choices easier (like using your reusable bag, remembering your reusable coffee mug, and taking the train).

There's a lot of science, information, and literature on developing habits. A habit is a regular tendency or ritual that is automatic to perform. For example, brushing your teeth in the morning is probably so automatic that you don't even *think about it;* you just *do it.* Feeding our dog every morning is another good habit of mine, even though our black Lab, Vicky, definitely wouldn't let me forget to feed her! Unfortunately, it's such a strong habit that even though I was supposed to have her fasted for an appointment one morning, I forgot and fed her anyway. Oops! Performing habits, instead of having to think deeply about these actions, saves brain energy, according to the experts. To develop new habits, use the "cue, routine, reward" method commonly referred to in habit-based research (like in the bestselling book *The Power of Habit* by Charles Duhigg). This system is about creating a cue (a trigger to get your brain into automatic mode), developing a routine (either physical or emotional), and following it up with a reward (to make it worthwhile for your brain to remember).[1] The reward in my case is a happy (and living) dog! If you're excited about developing a scientific approach to your new eco-friendly habits, I recommend making a list of new habits you'd like to change to reduce your trash footprint.

When I transitioned to a low-waste lifestyle, I was motivated enough to make changes that I typically didn't forget my reusable bags, travel mug, or whatever transition it was I was making at the time. It didn't feel like a conscious effort to make these new habits, although that may be different for you. I was more often unprepared for something like stopping for groceries on the way home from an event. My solution in that type of situation was to get my groceries and carry everything out by hand or bring the cart right to my car so I could make the transfer directly instead of taking a disposable plastic bag. Instead of falling back into the habit of getting the disposable plastic bag, I just found an alternative option that enabled me to still skip the bag. The choice was mine, and you'll be in situations too where you'll need to decide whether your values outweigh your convenience.

Also, a great tip is to make your new habits easier on yourself. If you tend to forget your reusable bags and your travel mug, leave them near your house keys, so that you see them when you leave home for the day. It's the perfect visual cue that you need to help you develop the habit! And while you're at it, cut the crap. There are no excuses here. If you are repeatedly saying, "Oh, I forgot my reusable bags," then come to terms with the fact that it's just not that important to you to bring them along! Move on and focus on *what is important to you* to change. Focus on what you're interested in, create the cues that work for you, and make the changes you want to make.

I believe in you. If you've read this whole book and made it to this last chapter to read these final paragraphs, then you're in it for the long haul! You've got passion and desire, and I believe those are the ingredients we need to make less-trashy-lifestyle changes and the fuel necessary to build new habits and to create everlasting change.

Reflection and Checklist

Chapter 12 is as much about reflecting on the work you've accomplished over your low-waste transformation as it is about moving forward with your less trashy life. I have no doubt that you'll be successful in having a lighter and less wasteful footprint on this planet if you implement even a few of the suggestions from this book. This month is a chance to reflect on what you've learned so you can decide what you want to change permanently in your life and take with you into your future. It's not just about your future, though, is it? If you're reading this book, it's probably also because you care about the future of this planet, with or without you. Maybe you want to leave this world better than you found it?

Choose your path forward and remember that it's not just about *you*. While I preach this approach of "you do you," that attitude is also very individualistic. You have to do you, because I don't know what your circumstances are. I'm not sure what level of income you have, what resources are available to you, or what barriers you may face. So when I say, "you do you," I simply mean, do what you can in your personal context and based on your personal preferences. May I suggest, however, that you keep in mind the broader community, the state of the planet, and future generations. At the end of this yearlong journey, here are some points to consider as you look forward to your less trashy future:

- Reflect on the changes you want to make and what changes don't work for you
- Define what sustainability looks like in your life, given your personal circumstances and preferences
- Develop the habits you need to live a less trashy life and a life filled with more joy
- Decide what legacy you want to leave behind

We all have a legacy, and it's up to each of us to decide what we leave behind. What will your legacy be?

We do not inherit the earth from our ancestors; we borrow it from our children.

—ORIGIN UNKNOWN

Afterword

Congratulations on making it through the whole book and undertaking the Don't Be Trashy Challenge! You've gone through tremendous changes, and I wish you all the best in continuing your journey toward a less trashy and more sustainable lifestyle. To help guide you beyond the pages of this book, I created Ten Sustainable Lifestyle Principles to live by.

Ten Sustainable Lifestyle Principles

1. Remember the three *R*s: reduce, reuse, and recycle
2. Add five more *R*s: rot, refuse, repair, repurpose, and rethink
3. Consume less stuff
4. Choose quality over quantity
5. Eat more plants
6. Live, learn, and grow
7. Be open-minded; don't judge (yourself or others)

8. Aim for progress, not perfection
9. Lead by example
10. Have fun

Many people feel overwhelmed by the environmental problems that we face worldwide. Rightly so! The problem is that overwhelm often leads to inaction. And even when we want to take action, the path forward for us as individuals isn't quite clear. This book is intended to be a road map to help everyday people take steps toward living lighter on this planet (kudos to you for taking action!). I recommend revisiting these Ten Sustainable Lifestyle Principles on a regular basis (I know I will); better yet, write them down, hang them up on your wall, or keep the list on your phone. Implementing these ten principles alone will make your life instantly less trashy. They may even provide inspiration for you to create your own mission statement or mantra.

Acknowledgments

Thank you to my husband for walking with me through this wild ride called life. You've been incredibly supportive of my journey toward entrepreneurship, and you've been by my side—with love, feedback, and snacks—every step of the way as I wrote this book. And you were with me that day snorkeling in Bali, so you saw where all this began!

Thank you to my family. To my parents, you have always been encouraging of my wildest dreams and of taking the path less traveled. I wouldn't be where I am today without all that love and support! To my sister, Katie, you're not just family; you're also my best friend, cheerleader, and sounding board. You've truly been a rock in my life over the years, laughing together during the good times and crying together during the tough times; thank you for helping me find my way when I needed it. And to my in-laws, Pat and LJ, thank you for your love, support, and regular check-ins; I'm so lucky to have you in my life!

Thank you to my editors. Laura, thank you so much for advocating for this project to come to life! And to Donna, thank you for believing in this book, getting on board, and helping create a powerhouse team. You both worked your magic! And thank you to Penguin Random House; the opportunity to become one of your published authors is truly an honor.

To my dearest friends, I'm very appreciative of the times you've touched base to see how things were going as I wrote this book—your daily encouragement, thoughts, and feedback never went unnoticed. That type of fuel kept me going!

To my online community, without you, this book literally would not have been possible! Thanks for taking the time to read my blog, *The Zero Waste Collective,* for following along on social media, and for connecting with me over DM, by email, and IRL.

And thank you to my clients. Without you, I wouldn't have taken the leap from a nine-to-five job to becoming an entrepreneur! I've had the chance to work with some of the most amazing brands and business owners, so thank you for the opportunity to work together as we try to make this world a better place.

Finally, I'd also like to acknowledge my amazing team of co-workers from back when I worked nine to five. You were all so supportive when I tossed around the idea of creating a green team, and later on you were encouraging in my final days at the office before I went out into the world of self-employment. You have no idea how much that meant to me. When I left, I mentioned that I would be writing a book; well, finally, here it is!

Notes

Preface

1. Ocean Conservancy, "Stemming the Tide: Land-Based Strategies for a Plastic-Free Ocean," McKinsey Center for Business and Environment, September 2015, https://oceanconservancy.org/wp-content/uploads/2017/04/full-report-stemming-the.pdf, p. 3.
2. Aarushi Jain, "Trash Trade Wars: Southeast Asia's Problem with the World's Waste," Council on Foreign Relations, May 8, 2020, https://www.cfr.org/in-brief/trash-trade-wars-southeast-asias-problem-worlds-waste.

Chapter 1: Trash Talk Basics

1. Zero Waste International Alliance, "Zero Waste Definition," http://zwia.org/zero-waste-definition/.
2. UN Environment Programme, "UN Report: Time to Seize Opportunity, Tackle Challenge of E-waste," January 24, 2019, https://www.unenvironment.org/news-and-stories/press-release/un-report-time-seize-opportunity-tackle-challenge-e-waste.

Chapter 2: Decluttering and Minimalism

1. Annie Leonard, *The Story of Stuff: The Impact of Overconsumption on the Planet, Our Communities, and Our Health—and How We Can Make It Better* (New York: Free Press, 2010), pp. 45–51.

2. Suzy Strutner, "Here's What Goodwill Actually Does with Your Donated Clothes," Huffpost, September 28, 2016, updated January 5, 2021, https://www.huffington post.ca/entry/what-does-goodwill-do-with-your-clothes_n_57e06b96e4b0071 a6e092352.

3. Joshua Fields Millburn, "Everything Is 100% Off If You Don't Buy It," The Minimalists, https://www.theminimalists.com/dont/.

Chapter 3: Conscious Consumption

1. Jenna Gavigan et al., "Synthetic Microfiber Emissions to Land Rival Those to Waterbodies and Are Growing," *PLoS ONE,* September 16, 2020, https://journals.plos .org/plosone/article?id=10.1371/journal.pone.0237839#sec016.

2. Sarah Gibbens, "You Eat Thousands of Bits of Plastic Every Year," *National Geographic,* June 5, 2019, https://www.nationalgeographic.com/environment/2019/06 /you-eat-thousands-of-bits-of-plastic-every-year/.

3. Stephen Leahy, "Microplastics Are Raining Down from the Sky," *National Geographic,* April 15, 2019, https://www.nationalgeographic.com/environment/2019/04/micro plastics-pollution-falls-from-air-even-mountains/.

4. Certified B Corporation, "About B Corps," https://bcorporation.net/about-b-corps.

5. Environmental Working Group, "About Us," https://www.ewg.org/about-us.

6. Adria Vasil, " 'It's Pretty Staggering': Returned Online Purchases Often Sent to Landfill, Journalist's Research Reveals," CBC, December 12, 2019, https://www.cbc.ca /radio/thecurrent/the-current-for-dec-12-2019-1.5393783/it-s-pretty-staggering -returned-online-purchases-often-sent-to-landfill-journalist-s-research-reveals -1.5393806.

Chapter 4: Pantry Goals

1. Michael Pollan, "Unhappy Meals," *New York Times Magazine,* January 28, 2007, https://www.nytimes.com/2007/01/28/magazine/28nutritionism.t.html.

2. Jonathan Safran Foer, *We Are the Weather: Saving the Planet Begins at Breakfast* (New York: Farrar, Straus and Giroux, 2019), p. 150.

Chapter 5: All Things Bathroom and Cleaning

1. Liza Torborg, "Mayo Clinic Q and A: Cleaning Products and Lung Health," Mayo Clinic, May 25, 2018, https://newsnetwork.mayoclinic.org/discussion/mayo-clinic -q-and-a-cleaning-products-and-lung-health/.

2. Tasha Stoiber, "What Are Parabens, and Why Don't They Belong in Cosmetics?," Environmental Working Group, April 19, 2019, https://www.ewg.org/california cosmetics/parabens#:~:text=The%2520concern%2520with%2520these%2520 chemicals,can%2520also%2520cause%2520skin%2520irritation.

3. Environmental Working Group, "Top Tips for Safer Products," https://www.ewg .org/skindeep/contents/top-tips/.
4. Environmental Working Group, "EWG's Healthy Living: Home Guide," https:// www.ewg.org/healthyhomeguide/.
5. Environmental Working Group, "EWG's Healthy Living: Home Guide: Cleaners & Air Fresheners," https://www.ewg.org/healthyhomeguide/cleaners-and-air -fresheners/.
6. Environmental Working Group, "EWG's Healthy Living: Home Guide: Cleaners & Air Fresheners."
7. David Suzuki Foundation, "Does Vinegar Kill Germs?," https://davidsuzuki.org /queen-of-green/does-vinegar-kill-germs/.

Chapter 6: Outfit Repeater

1. Kelly McSweeney, "This Is Your Brain on Instagram: Effects of Social Media on the Brain," Now. Powered by Northrop Grumman, March 17, 2019, https://now .northropgrumman.com/this-is-your-brain-on-instagram-effects-of-social-media -on-the-brain/.
2. Value Village, "Rethink Reuse," https://www.valuevillage.com/thrift-proud/rethink -reuse/.
3. Emma Johnson, "The Real Cost of Your Shopping Habits," *Forbes,* January 15, 2015, https://www.forbes.com/sites/emmajohnson/2015/01/15/the-real-cost-of-your -shopping-habits/?sh=903519c1452d.
4. Elizabeth Cline, "The History of a Cheap Dress," Etsy Journal, May 31, 2011, https://blog.etsy.com/en/the-history-of-a-cheap-dress/.
5. Nicole Caldwell, "How to Build a Capsule Wardrobe, According to the Woman Who Invented It," Green Matters, May 24, 2019, https://www.greenmatters.com/p /capsule-wardrobe-invention.
6. Hubbub, "Encouraging Sustainable Choices by Providing an Alternative to Black Friday Sales," https://www.hubbub.org.uk/brightfriday.

Chapter 7: The Subtle Art of Refusal

1. Erika Giovanetti, "Holiday Debt Averaged $1,381 in 2020, Reaching a 6-Year High amid Pandemic," Magnify Money, December 28, 2020, https://www.magnifymoney .com/blog/news/2020-holiday-debt-survey/.
2. Stanford University, "Frequently Asked Questions: Holiday Waste Prevention," https://lbre.stanford.edu/pssistanford-recycling/frequently-asked-questions /frequently-asked-questions-holiday-waste-prevention.

Chapter 8: Family and Friends

1. Sid Perkins, "Why Do Seabirds Eat Plastic? They Think It Smells Tasty," *Science*, November 9, 2016, https://www.sciencemag.org/news/2016/11/why-do-seabirds-eat -plastic-they-think-it-smells-tasty.
2. "Only 9% of the World's Plastic Is Recycled," *Economist,* March 6, 2018, https:// www.economist.com/graphic-detail/2018/03/06/only-9-of-the-worlds-plastic-is -recycled.
3. Vanessa Forti et al., "The Global E-waste Monitor 2020: Quantities, Flows and the Circular Economy Potential," United Nations University and United Nations Insti- tute for Training and Research, SCYCLE Programme, International Telecommunica- tion Union & International Solid Waste Association, Bonn/Geneva/Rotterdam, p. 9, https://globalewaste.org/.
4. UN Environment, "Worldwide Food Waste," https://www.unenvironment.org /thinkeatsave/get-informed/worldwide-food-waste.
5. UN Environment, "Worldwide Food Waste."
6. Plastic Soup Foundation, "The World's Population Consumes 1 Million Plastic Bot- tles Every Minute," https://www.plasticsoupfoundation.org/en/2017/07/the-worlds -population-consumes-1-million-plastic-bottles-every-minute/.
7. Mina Sinai, "How Many Times Can Recyclables Be Recycled?" Recycle Nation, June 27, 2017, https://recyclenation.com/2017/06/how-many-times-can-recyclables-be -recycled/.
8. Lilly Sedaghat, "7 Things You Didn't Know About Plastic (and Recycling)," *National Geographic,* April 4, 2018, https://blog.nationalgeographic.org/2018/04/04/7-things -you-didnt-know-about-plastic-and-recycling/.

Chapter 10: Money Matters

1. "North Face Widow Tompkins Donates Land for Chile Parks," BBC News, March 16, 2017, https://www.bbc.com/news/world-latin-america-39292600.
2. Tompkins Conservation, "About Kris and Doug Tompkins," http://www.tompkins conservation.com/about_kris_and_doug_tompkins.htm.

Chapter 12: Wrap It Up

1. Charles Duhigg, *The Power of Habit: Why We Do What We Do in Life and Business* (New York: Random House Trade Paperbacks, 2014).

Index

About the Author

Tara McKenna launched *The Zero Waste Collective* after working as an environmental planner and witnessing over-flow trash intermingling with fish and coral in Southeast Asia. This online community offers simple strategies to live more sustainably and aims to inspire and empower people from all walks of life to find more joy by living with less waste!